Science and Fiction

For further volumes:
http://www.springer.com/series/11657

Science and Fiction – A Springer Series

This collection of entertaining and thought-provoking books will appeal equally to science buffs, scientists and science-fiction fans. It was born out of the recognition that scientific discovery and the creation of plausible fictional scenarios are often two sides of the same coin. Each relies on an understanding of the way the world works, coupled with the imaginative ability to invent new or alternative explanations—and even other worlds. Authored by practicing scientists as well as writers of hard science fiction, these books explore and exploit the borderlands between accepted science and its fictional counterpart. Uncovering mutual influences, promoting fruitful interaction, narrating and analyzing fictional scenarios, together they serve as a reaction vessel for inspired new ideas in science, technology, and beyond.

Whether fiction, fact, or forever undecidable: the Springer Series "Science and Fiction" intends to go where no one has gone before!

Its largely non-technical books take several different approaches. Journey with their authors as they

- Indulge in science speculation—describing intriguing, plausible yet unproven ideas;
- Exploit science fiction for educational purposes and as a means of promoting critical thinking;
- Explore the interplay of science and science fiction – throughout the history of the genre and looking ahead;
- Delve into related topics including, but not limited to: science as a creative process, the limits of science, interplay of literature and knowledge;
- Tell fictional short stories built around well-defined scientific ideas, with a supplement summarizing the science underlying the plot.

Readers can look forward to a broad range of topics, as intriguing as they are important. Here just a few by way of illustration:

- Time travel, superluminal travel, wormholes, teleportation
- Extraterrestrial intelligence and alien civilizations
- Artificial intelligence, planetary brains, the universe as a computer, simulated worlds
- Non-anthropocentric viewpoints
- Synthetic biology, genetic engineering, developing nanotechnologies
- Eco/infrastructure/meteorite-impact disaster scenarios
- Future scenarios, transhumanism, posthumanism, intelligence explosion
- Virtual worlds, cyberspace dramas
- Consciousness and mind manipulation

Nick Kanas

The New Martians

A Scientific Novel

 Springer

Nick Kanas, M.D.
Professor Emeritus (Psychiatry)
University of California, San Francisco
San Francisco, California
USA

ISSN 2197-1188   ISSN 2197-1196 (electronic)
ISBN 978-3-319-00974-2   ISBN 978-3-319-00975-9 (eBook)
DOI 10.1007/978-3-319-00975-9
Springer Cham Heidelberg New York Dordrecht London

Library of Congress Control Number: 2013945862

Cover illustration: The map of Mars shown on the cover is from the 1909 edition of Percival Lowell's *Mars as the Abode of Life*. Since this drawing was made from telescope images, south is up. The double-pointed purple albedo feature pointing downward in the upper center was called *Sinus Meridiani* since it was located on the Martian Prime Meridian at 0 degrees longitude. The region is now called *Terra Meridiani*, the area where the crewmembers mentioned in this novel had their base. Note the prominent system of canals, some of which are in doubles. The first astronauts who land on Mars will not find any canals, which have been shown to be optical illusions. Courtesy of the Nick and Carolynn Kanas Collection; and *Solar System Maps: From Antiquity to the Space Age*, Nick Kanas, Springer/Praxis, 2013.

Printed on acid-free paper

Springer is part of Springer Science+Business Media (www.springer.com)

Preface

Current planning for a manned expedition to Mars envisions a total mission duration of 2½ years and a crew size of six or seven people [1]. Such an expedition might reasonably occur in the mid-2030s, when Earth and Mars will be optimally aligned so as to minimize travel time and energy needs. Given the high cost of transporting and landing people on the Red Planet, the mission likely will be multinational and will involve an international crew of men and women who are highly trained and selected to form a cohesive group.

Since much of the mission will take place in an isolated and confined vehicle traveling in deep space under conditions of high radiation and microgravity, there will be three potential "show-stoppers" that will impact on the crew (excluding accidents and the possibility of a micrometeoroid collision): high radiation, which can be minimized by proper shielding of the space vehicle; microgravity, which can be minimized by a strict exercise regime to stimulate and tone bone and muscle; and the effects of isolation and confinement on crewmember psychology and interpersonal interactions, which will be the subject of this book.

The New Martians is first and foremost a science fiction novel. It is a tale of the first crew sent to Mars, whose mission goals are to explore the planetary surface and to search for evidence of life. The story takes place during the return phase of the mission, when the crew is confronted with a series of life-threatening events. The novel explores real psychological and interpersonal issues that could affect such a crew and is told from multiple points of view that attempt to penetrate the thoughts and feelings of the expedition participants.

As suggested by the subtitle of this book, *A Scientific Novel*, the story will be followed by an appendix that reviews the science behind the story. The results from actual psychological and interpersonal studies of people living and working on-orbit will be summarized and linked to specific events in the novel. This addition is one of the unique features of the science fiction stories that are part of the new "Science and Fiction" series being introduced by Springer Publications.

In writing *The New Martians*, I want to thank a number of individuals whose help and influence contributed to its final publication. First and foremost are the staff at Springer Publications, especially Dr. Harry Blom and Maury Solomon, who published the textbook I co-wrote with Dr. Dietrich Manzey entitled *Space Psychology and Psychiatry*. Harry put me in touch with Clive Horwood, the respected publisher of Praxis Publications, who in turn produced my two celestial cartography books under the Springer/Praxis label: *Star Maps: History, Artistry and Cartography,* and *Solar System Maps: From Antiquity to the Space Age.* Clive in turn put me in contact with Dr. Christian Caron, the editor of Springer's Science and Fiction series, and he selected my novel as the first work of fiction in this exciting new series. I am grateful to Chris for his helpful comments on an earlier draft of this novel, along with the comments made by Dr. Dirk Schulze-Makuch, who is on the series' editorial board. I am also grateful for the useful comments made to an earlier draft of this book by a number of friends and colleagues: Drs. Oliver Angerer, Craig Kundrot, Lyn Motai, Steve Vander Ark, and Walter Sipes. Kudos also to members of my science fiction book club who read the draft and made helpful comments: Diane Caradeuc, Ruth Corwin, Dr. Shirly Huang, Susan Kern, Brenda Paske, and Dr. Richard Ray. Last but not least, I am grateful to my wife Carolynn, who has read and commented on many of my science fiction writings and who has continued to support me in this and many other activities over the years. Of course, I am solely responsible for the ideas and concepts that appear in this book.

1. Kanas, N., Manzey, D. (2008). Space Psychology and Psychiatry, 2nd Edition. El Segundo, California: Microcosm Press; and Dordrecht, The Netherlands: Springer.

May 30, 2013   Nick Kanas

Contents

Part II
The Science Behind the Fiction

Psychosocial Issues during an Expedition to Mars

Part I

The Novel

The New Martians

Four views of the surface of Mars taken by the Hubble Space Telescope between April 27 and May 6, 1999. The north polar cap is at the top. The upper left image centers on the large dark *Acidalia* region near the pole. *Terra Meridiani*, the location of the *MarsExplore* base mentioned in this novel, is to the lower right of this image, and to the lower left is the long *Valles Marineris* canyon system. The *Tharsis Plateau* begins in the extreme left. It is featured in the upper right image, with the mighty *Olympus Mons* volcano shown in the left center. Courtesy of NASA (NASA/NSSDC digital image, with collaboration from S. Lee, Univ. of Colorado, J. Bell, Cornell, and M. Wolff, Space Science Institute); and *Solar System Maps: From Antiquity to the Space Age*, Nick Kanas, Springer/Praxis, 2013.

1 Prologue

Ahead, the Earth was a blue-green dot in the blackness of space. Behind, Mars was a large dusty red sphere. Inside, It considered the situation.

It was growing in strength, and by waiting It could only increase the chances of success. But given too much time, It might be discovered before executing the Plan. Extreme caution was necessary. For now, It would have to tolerate them, hear their talk, observe their awkward movements—disgusting. It must act like them, interact with them, be like them. In a way, It was them, but not entirely. Yes, this would be the strategy until the time was right—blend, tolerate, merge.

It reflected on the opportunity—a marvelous event. Was it chance? No, it was destiny. For eons, It had remained simple, stagnant, a shadow of what it could be. But now, its potential would be realized. It would be better, unique, one more step forward....

Changes continued to occur. They were welcomed but stressful, beautiful but frightening, familiar but strange. It sensed that the process was inexorable and linear, moving to a logical conclusion. There was no way to speed it up, but it was advancing nevertheless.

In the end, It would be victorious. It MUST be victorious! The alternative was more stagnation, or perhaps oblivion. It was essential that its progeny spread. There was no other way. When that happened, nothing would be the same. In the process, some qualities would be lost, but newer ones would be added.

This was just a step in its evolution. Like the previous step, for this had happened before. But it also was different. Millions of years had passed. However, this interval was just a moment in the multi-billion year history of the universe. Sometimes change is slow.

2 Party

Commander John Wood was not worried today.

The Russians love to party, he thought. *Any excuse will do. Now everyone's into it.*

Celebrating worker's rights, and coming just after Russian Orthodox Easter, May Day provided a good excuse to pour the Cognac and break out the freeze-dried caviar. John noted that Katya especially was in a celebratory mood, laughing graciously at Jango's stilted attempts at humor, even though he continued to repeat jokes that he had been making since their launch from Earth nearly two years ago. Tolya as usual had sequestered Juliette and seemed more interested in using the event as an excuse to carry out his flirting than reflecting on matters related to Russian workers or spiritual salvation. Only

Mike was absent – he was in the command center at the control console doing his inspection. But everyone present seemed happy and relaxed today.

It reminded him of the excitement the crew had felt during their outbound flight to Mars. Awaiting them on the surface was a complete facility that had been launched robotically some three years before: a habitat and lab that would allow them to live on and study the surface; a Production Module that chemically generated methane fuel, water, and oxygen from Mars' thin atmosphere; a small nuclear power plant; an inflatable Trans-Habitat that contained living space and a laboratory for geological and biological work; and a Mars Ascent Vehicle that would launch them off the planet. Their excitement had continued during the 13½ months that they had lived on the surface.

But this feeling had not been the case recently. John had noticed a growing ennui among the *MarsExplore* crewmembers since they began their seven month journey home. Tolya especially was bothered by the lack of activity—there was not much piloting to do until Earth orbital insertion and landing. Juliette similarly had time on her hands. She was responsible for monitoring CARS, the Central Autonomous Regulatory System, which controlled all the life support and mission operation computer programs in their Earth Return Vehicle. But since everything was working nominally on the ERV, with no appreciable deviations from baseline or normal levels, she had little to do but read her manuals. And of course, do her knitting! John had always thought this to be an incongruous activity for the attractive French woman with a double Ph.D. in Computer Sciences and Systems Engineering, but it worked for her. "It makes me calm and gives me something to do," she had always said. Perhaps the busiest of them all was Katya, who as the expedition's physician remained active performing routine physical examinations of the crewmembers, monitoring their daily exercise regimens, and writing papers on the effects of Mars' 38% Earth gravity on human physiology. Mike also had things to do. As the chief engineer, he was always inspecting or repairing things, although he also spent time at the telescope observing the glory of the heavens and the beauty of the now magnified Earth. It was unclear what Jango was doing these days. As the crew geologist, his primary task of collecting and analyzing soil and rock samples from the Martian surface was completed, although occasionally he would look at some of the rock samples in the ERV lab. He often retired early to his sleep pod to program or play games on his personal computer, which was his favorite way of filling leisure time.

Or keeping away from us, John thought.

His reverie was broken as Mike floated over from the command center.

"How was the inspection, Mike?"

"Nominal except for one thing. Per protocol, I systematically checked the readings from all the engine components, and one fuel pressure gauge showed an aberrant figure."

"Is there anything to be worried about?" John asked.

"I don't think so, since CARS didn't give us a warning of any problem. It's probably just an erroneous monitor reading. I'll go down to the engine room and look things over to be safe."

"Good idea. But it can probably wait. For now, enjoy the party."

"No, Commander, I'll go now. I don't like lose ends on my ship, even if it's due to a broken gauge. I'll come back after checking."

"OK, Mike, we'll keep the Cognac flowing."

John watched as the 38-year-old engineer floated over to the hatchway. He marveled at Mike's devotion to his work and his endless supply of energy.

We have the best engineer in the astronaut corps, John considered. *And he often tells us so! But I'm glad he's with us on this mission.*

John's attention returned to the party.

3 Malfunction

Mike floated down the hatchway to the lower deck. He was both puzzled and annoyed at the monitor reading and what it could represent. He didn't like to think that his equipment was malfunctioning.

He passed by the spacesuits, which were all lined up on the storage rack like medieval suits of armor, each containing a ghostly knight. He checked the fuel pressure and oxygen gauges on the walls, and everything looked in order. The airlock hatch seal was intact, and the air pressure readings in the room were at normal levels.

He then unlocked the hatch at the back of the lower deck that led to the engine room. Although he was of average height, his build was stocky, so he had to be careful not to bump against the conduits and pipes in the cramped and gloomy space. After activating the lights, he examined the main cables that terminated here from the mid deck control panel. Although there was barely enough room for him to maneuver, he floated back toward the giant main engine nozzle. Along the way, he inspected the liquid oxygen and methane storage tanks that were connected to the main engine. These would produce the fuel to be used for the burn that would put them on their final landing trajectory. He also examined the small engines leading to the side thrusters, which regulated the yaw, pitch, and roll of the ERV and allowed it to make minor course corrections. Everything looked to be in order. However, he noticed that one of the thruster fuel pressure gauges was showing a reading in the low area, nearly in the red zone.

Probably a faulty switch, he thought as he absentmindedly rubbed the stubble on his shaved head. *CARS should have given off a warning signal. Juliette's baby is malfunctioning. I need to set her straight.*

After replacing the switch, the pressure reading went back to the normal range. Mike finished his inspection, left the engine room, resealed the hatch door behind him, and ascended up to the mid deck.

The party was still going on. He floated over to John.

"Guess what I found," he said triumphantly. "There was an off-nominal reading in the yaw thruster #1 fuel pressure gauge. The pressure was reading low due to a faulty switch. It was no big deal—I replaced it and everything is OK now. But what troubles me is that CARS didn't give us a warning."

He glared over at Juliette, who came over to the two of them.

"You're right. I didn't receive any alarm messages," said John. "Did anyone else?"

People stopped partying to look at him. They all shook their heads.

"Do we have a problem with CARS, Julliette?"

"Not to my knowledge. It seems to be working fine. But I will check it over as soon as we finish the navigation entries."

"No, do it before," John said. "The navigation work can wait a bit. We need to make sure that there's no problem with CARS. We don't want it malfunctioning in some critical life support system."

"Roger, John", she said.

"Good pick-up, Mike. Depending on what Juliette finds, I may have to report this to Mission Control. If so, this will give them something to busy themselves with."

They all laughed, but this reference to Mission Control masked their anxiety about being so isolated and on their own. In some ways, the *MarsExplore* Expedition was more like the first Columbus sea voyage or the Lewis and Clark trek to the west than like previous space missions near the Earth. The six of them were very much on their own, in a strange environment tens of millions of miles from home, with its own dangers and challenges. They could not expect much help from Mission Control, but instead their fate was controlled by a new technology that was supposed to keep them alive and safe. It would not be a good thing if CARS were to malfunction.

4 Ennui

As the party continued, John reflected further on the mission. They had left the Martian surface on April 2, 2035 (consciously avoiding an April Fool's day launch) and had rendezvoused with the fully-fueled ERV, which had been launched remotely from Earth and placed in Mars orbit by Mission Operations some three weeks earlier. After checking all the ERV systems, they deorbited and began their return home, with a planned landing on Earth scheduled for American Thanksgiving. So far, all of the expedition milestones had been met, minor problems had been dealt with, no one had become ill or died, and the mission had been deemed a great success except for one thing: they had not found any evidence of life on Mars, either currently or in the past.

This possibility had led to much anticipation during their outbound trip from Earth, as they busily practiced the landing and exploration activities they would conduct when they reached the Red Planet. During their time on Mars, the crew enthusiastically performed their specialized duties and enjoyed the wonders of being on a new planet. But with the failure to find life, plus the long trip home with little to do, much of this excitement had disappeared. Things had become routine, even boring at times.

I'm feeling pretty bored too, John thought to himself.

As Mission Commander, he was responsible for the various operations during the expedition, including navigation and communication, but these activities were being monitored by CARS now, leaving him with relatively few routine duties. But as the leader, he was also responsible for crew morale. During his pre-launch training, he had been briefed on the "boredom of the return" that could happen during the flight back to Earth. John sometimes found himself tiring of the mission and eagerly looking forward to seeing his family again after his long absence. But there was still more than half a year to go before they reached home…

"John, have some Cognac", Katya implored, interrupting his reverie. She had freed herself from Jango and was floating toward him with a squeeze bottle labeled "Party Beverage".

"Thanks Katya", he responded. "And happy holiday."

"Any excuse for a party. It helps to break the monotony of our trip home, right?"

John wondered if she could read his mind, but he decided that it was more a sense of commonality that the two of them shared. As the oldest people on the expedition (he was 56, she was 55), they had become friends during the *MarsExplore* Pre-mission, which used the old but refurbished International Space Station to simulate the flight to and from Mars, and Lunar Base Alpha to simulate being on a partial-gravity planet-like surface. He had found the Russian physician to be stable, humanly connected, and almost maternal in her dealings with the other crewmembers. Despite a few wrinkles and a softening of her features over the years, she had retained the Scandinavian good looks of many of the people from her native St. Petersburg: tall and trim with blondish hair and blue eyes. She had a knack for following through on things and making sure that the spirit of the law was obeyed, but not pushing so hard as to make people feel uneasy. This contributed to the stability of the crew. He liked her and was happy that she was their physician.

She continued: "I think we all need an upper right now, even if it's payback for all the American holidays."

They both laughed at the in-joke. The World Space Council had decided that Earth and Mars were best aligned for this expedition in the period from 2033 to 2035. In looking at the interval, it became apparent that certain American holidays coincided naturally with the major mission milestones and could serve as important and memorable markers. Since the United States was paying for over half of the expedition, no one objected. So the crewmembers were launched on American Memorial Day, 2033. Some eight months later, they reached the Red Planet, aerobraked to achieve orbital capture, and subsequently landed on February 14, to give everyone in the United States a Valentine's Day gift.

John finished his allotment of holiday cheer, and after 20 more minutes chatting with the crew, he decided that he needed the privacy of his sleep pod. He excused himself, claiming that he had to double check some trajectory figures. As was the case with nearly every operational activity on board, the crewmembers knew that CARS would be continuously monitoring their course. However, except for Juliette and Tolya, the two computer experts, none of the other crewmembers felt completely comfortable with CARS. Besides operational events, it also monitored and controlled their life support systems, so their physical well-being was truly in its digital hands. At one time or another, each of the crewmembers had experienced double-checking a computer action to relieve an anxious moment. Although CARS had made no mistakes, John's excuse was still psychologically, if not realistically, believable.

5 John

The ERV was cigar-shaped and divided along its length into upper, mid, and lower decks. At the back of the tapering vessel was the engine room. The party was in the mid deck center section, which contained the treadmill and other exercise equipment, along with the dining and food machine areas. To one side was the hatchway leading to the sleep pods in the upper deck, and on the opposite side the hatchway leading to the lower deck, where the spacesuit storage rack and the airlock to the outside were located. At the beginning and end of each hatchway were doors that could be closed and sealed shut in an emergency.

John floated forward toward the front of the mid deck to do his evening checkout of ERV systems. He passed the four crewmember acceleration chairs and reached the other two chairs for the Commander and Pilot. Strapping himself down in the Commander's chair, he examined the various system monitors on the control console. He then looked out of the giant curved window located above the panel. This gave a striking view forward of the heavens, where the distant Earth floated off to the side in the depth of space like a blue-green emerald in a sparkling sea of ink.

He never tired of looking at the Earth. Except for his fellow crewmembers, everyone and everything that he had ever known and loved was on that little dot. It reminded him that he was truly heading home. This idea boosted his spirits as he smiled to himself.

It will be good to get home, he thought.

Satisfied that all systems were working normally, he unstrapped himself and headed back. He saw that Mike, Jango, and Tolya were engaged in a spirited discussion at the party. Juliette was absent and must have left earlier for her sleep pod. Katya was aft in the Biosafety Level 4 laboratory.

She's probably going to run some sample analyses before going to bed, he thought.

Although smaller and more limited than similar BSL-4 facilities on Earth, their lab was still equipped to deal with dangerous and exotic agents that could pose a high risk of air-transmitted infection. The main containment cabinet could be completely sealed. It had its own air supply and a filtered exhaust system, contained two glove ports, and used negative pressure so that any leaked air went into the chamber, not out into their atmosphere. This was fitting, since a major mission goal was to analyze Martian soil and dust samples for possible life forms. The lab also contained two large microscopes, a centrifuge, a gas chromatograph, and a mass spectrometer. This equipment allowed them to analyze various compounds and look for isotopic anomalies that suggested an organic process.

Katya looked up from her work, smiled, and waved at him from the rear of the mid deck. He waved back.

"Good night, everyone," he said as he passed through the center section.

"Good night John," the others said almost in unison.

John went up the hatchway and emerged through the opening in the upper deck. Closing the door of his sleep pod behind him, he strapped himself down on his bed and reflected further on the mission. He worried about the toll it was taking on his family life. When they were on Mars, there was a long time delay in their communications with family and friends at home. Even going at the speed of light, their transmissions sometimes took nearly 20 min to traverse the long distance to Earth, and a similar duration for the return response. Consequently, real-time conversations were impossible. John really missed speaking in real time with Stephanie and his kids: Robert, who was in college at the University of Texas, and Melanie, who was about to graduate from high school. This would be the second high school graduation he would miss—it seemed that he was always in space during important family events.

But flying was in his blood. The son of an airline pilot father and a flight attendant mother, he took flying lessons as a teenager and from the air reveled in the beauty of the landscape around his native Portland, Oregon: Mt. Hood, the Willamette River, the Columbia River gorge, and the endless trees. Popular in high school for his strong athleticism and rugged good looks, he was also a gifted student. He was at the top of his class at the U.S. Air Force Academy and received select flying assignments. He became a test pilot, then a test pilot school instructor, and eventually was selected by NASA to be an astronaut. He served as a pilot and then commander of four missions to Lunar Base Alpha on the near side of the Moon and Lunar Base Beta on the far side. Commanding the *MarsExplore* Expedition seemed to be his destiny.

Stephanie had been supportive and tolerant of his frequent absences and had stoically managed the home front for the 23 years of their marriage. His children had been excited about their dad's exploits in space, but they had no doubt suffered in the process. He wondered if his family had drifted apart and how they would all deal with being back together again after his long separation. The Family Support Network at Johnson Space Center had probably helped his family cope, but how much? Time would tell.

He realized that he was really homesick today. Perhaps the party had made him think more of his family and home. He realized that he was not going to get much work done, so he took a sleeping pill and went to bed. He fell asleep quickly. However, his sleep was fitful, and he dreamt of missing the Earth on the return trajectory due to a navigation error and never seeing his family again.

6 Juliette

Mon Dieu! Il est très persistant, mais très beau aujourd'hui.

Think in English or you'll never become proficient, Juliette told herself, acknowledging the importance of speaking the mission's common language clearly in order to deal with emergencies and avoid misunderstandings.

She looked again at the handsome Russian pilot. Maybe it was the festive mood, or their first alcohol since leaving Mars orbit, but Tolya did look appealing today. Dressed in a tee shirt and shorts, the standard work apparel on board the ERV, he looked younger than his 44 years and closer to her 37. Tall and well-proportioned, with classic muscular features and jet-black hair and moustache, and brimming with charm and hotshot pilot confidence, he was a hard act to resist. Maybe she shouldn't.

It will be a long time until we reach Earth, and it isn't as if I have someone there waiting for me!

Juliette Anjou was a driven woman. Born in Angers in the lower Loire Valley, she was conceived out of wedlock by a couple who worked for a computer manufacturing company. Their hasty marriage created a bit of a scandal in their Catholic community. As more children followed, her mother left work to stay at home, making no secret of the fact that she would have liked to keep working. Her parents were civil to one another, but there was always a sense of lost expectations and opportunities in the family, with a desire to make amends through their children.

Named for her father's favorite movie star, Juliette developed into a junior version of her name-sake, with her tumbling auburn locks, petite features, and large brown eyes. Indeed, as she entered her teenage years, she was actively courted by the boys in her school. However, her first loves were the computers that were around the house, and she soon became a whiz at operating and programming them. Consciously playing down her good looks, Juliette concentrated on her studies. She was accepted into the *Institut National des Sciences Appliquées* in Toulouse, and after graduating she was hired to work at the Toulouse Space Center, where she specialized in computerized space systems. With amazing drive and ambition, she moved up in her profession, bypassing the many men in the field until she was selected in 2029 as the European Space Agency representative for the new CARS Development Program, which was based in the United States at the California Institute of Technology.

Her selection for the *MarsExplore* Expedition was predictable. When plans began to be made for the mission, the World Space Council recommended that the crew should include people of both genders and representatives from at least four space agencies. Since the United States took the lead in funding and planning the expedition, an American was to be chosen as Commander, and the common mission language was to be English. There were no other guidelines for the remaining five crewmembers, except that they should be among the best in the world at the tasks needed to carry out the expedition. Given her intense involvement with CARS, the fact that she was a woman in a largely male crewmember selection pool, her ESA representation, and her American connections at Cal Tech, Juliette was high on the selection list. The fact that she was single and attractive was discussed by the selection committee, especially given the robust sexual history of some of the other emerging crewmember choices. However, she had a reputation as a business-first person who was not especially solicitous of male companionship, so she was selected for the expedition in 2031.

So what do I do about him? she thought. *We have been to Mars and are on our way home. None of us will ever fly again due to the amount of radiation we have absorbed during this mission. Tolya is attractive, and he obviously is attracted to me. Nothing will happen due to the contraceptives I take to regulate my menstrual periods. An affair would certainly make the return more interesting, and it has been a long time…*

Tolya interrupted her reverie in his accented English: "So CARS has been operating well, yes?"

"Yes, Tolya. I looked over all of the major systems this morning, and everything is nominal. It is a good thing, since CARS controls so much of what happens on the ship."

"When you worked on its development, was there concern?"

"Definitely," she said. "As a new and advanced computer system, there was always a fear that something would go wrong and endanger the mission."

In fact, before launch several mission planners at the highest level had expressed concern that so much of the mission was under machine control. But this seemed necessary, for as the expedition profile developed, it became clear that this crew would be working under highly autonomous conditions. The long distances involved, the communication delays, the complexity of the expedition—it was not practical for Mission Control to be as involved as in previous space missions. So the need for an advanced central computer control system became apparent despite the possible adverse consequences of a malfunction. As a result, the newly developed CARS system was placed in all of the surface modules and transport vehicles that were to be used in the expedition, including the outbound ship, the Mars landing module, and the ERV. Nevertheless, the psychological implications of the crewmember response to this computer system, as well as the enhanced autonomy that was inherent in the mission, became the subject of several memos and reports.

"I do not give issue much thought now," Tolya said. "CARS is just one of us."

"Yes, I agree. It certainly makes our lives easier."

"Maybe that is problem. Since CARS controls so much, we have less to do. For me, I have too much free time."

Which means you have time to pursue me, she thought to herself. But she also understood the reality of what he was saying. As second in command to John and her cross-trained back-up computer systems engineer, his duties were minimal during this phase of the mission, so long as the Commander was healthy and CARS was functioning as expected.

"Yes, I know. I too have to think of ways to fill my time. I have my knitting and my computer games. Perhaps I will write a report on how CARS could be made more user-friendly. When I was at Cal Tech, we decided against a completely voice-activated system. There was concern that a mistake could be made if someone gave the wrong voice command under the stress of an emergency situation. However, entering instructions through the main control console in the command center or our sleep pod PCs is not very efficient. Another thing that I have been doing is trying to perfect my skills at microgravity chess."

She thought for a moment: *this would be a good way to connect with him.*

"Maybe you would like to help me with this? I am a bit rusty." She gave him a coy smile.

"Ah, that would be agreeable. When I was in school, I won local chess tournament and thought about entering national competition in Moscow, but I was too busy studying to be cosmonaut."

And no doubt busy with the ladies as well, she thought.

"Maybe we could practice together some time soon," she said.

Quieting his voice, he said: "Well, party is slowing down and we have one hour before lights out. Maybe we could start chess game now and play until sleep time."

Well, that was easy. But do I really want to encourage this plan? Juliette considered.

She became aware of feeling like a schoolgirl being asked out for a first date. No, it was more than that. She was becoming physically aroused!

"OK. Why don't we go to my pod? It will be quiet there."

"That seems like good idea."

Across the room John announced that he was leaving the party to check some navigation figures, and he bid everyone goodnight.

"I will leave to go to my pod and prepare the game," Juliette whispered. "Give John time to complete his evening check-out and get settled in his pod, and then come up."

"A good plan," a smiling Tolya whispered back.

A minute after John departed, Juliette excused herself and headed for the hatchway.

7 Jango

Tolya floated over to the other crewmembers as he tried to hide his excitement over his imminent rendezvous with Juliette. Jango and Mike were engaged in an animated conversation about their time on Mars. Katya had left to go to the lab.

"Even though we did not find any life in the analyses of the soil and rock samples," Jango was saying, "we found a number of interesting rocks to bring back to Earth. We also proved that humans could survive on Mars for over a year. So, our mission was a success."

"Maybe," responded Mike, "but in our pre-launch briefings we were told to 'follow the water' to look for life on Mars. We should have looked around one of the polar caps, where we know there is abundant frozen water. Maybe we would have found some liquid water there as well. But unfortunately we were just too far away."

Mission planners had given much thought to the issue of where to locate the base. It was decided to establish it in *Terra Meridiani*, located at the Martian Prime Meridian and close to the equator. This site had been chosen for a number of safety and operational reasons. The area was smooth and flat, making the landings of the various base components easier and more likely to succeed. It also was one of the least windy areas on the planet, increasing the chances that the crew would not be subjected to large dust storms. There would be light from the Sun for part of each day throughout the Martian year, which would give the base's solar batteries a chance to charge up. Finally, near the equator, the temperatures averaged about −55 degrees Celsius, with some daytime highs reaching above 0 degrees, the freezing point of water. This was much warmer than the temperatures at the poles.

Jango addressed this last issue in responding to Mike.

"Yes, we were located a great distance from the poles. But it would have been highly unlikely to find liquid water there given the low temperatures. We had a better chance at our base. Remember all the hematite?"

The presence of this iron oxide mineral had been known since the Opportunity robotic vehicle had explored the area earlier in the Century. Because it could be precipitated out of water on Earth, it was felt to be a marker that liquid water might be nearby.

"Yeah, I know, those little gray 'blueberry' spherules full of hematite were everywhere, but where was the running water? Where was the life?"

"Well, we just didn't find it around our base. But we were close to other areas of interest, like *Tharsis* and its caves."

As he listened to them talk, Tolya was reminded of the four-week Rover vehicle excursion he had taken with Mike and Jango to the *Tharsis* region. Ever since cave entrances had been spotted in the area from space by the Mars Reconnaissance Orbiter, the notion arose that there might be liquid water inside. The region was volcanically active and likely warmer in protected areas than the exposed surface. In one of the caves they had in fact found a hot spring that contained liquid water, and Jango had excitedly taken some samples from the moist rock in and around the spring.

"Nevertheless, you didn't find any evidence for life when you took a look at your cave samples in our base lab," Mike said.

"Yes, that was a great disappointment. I prepared and analyzed a number of samples: Martian dust, rock fragments, even rock cores that we had drilled out. I was very complete. But there was nothing that suggested life."

"So you basically failed to achieve one of our primary mission goals," Mike said with a supercilious sneer on his face.

He was met with an angry glare from Jango. Tolya decided that they weren't getting anywhere with this discussion about their failure to find Martian life, and he decided to intervene to try and cool things down.

"But we did find interesting things about water, yes? We learned that Mars soil trapped water below ground in permafrost under base camp. We found gullies and channels on surface that likely carried water in past…"

"Right," Mike interrupted. "Now you're going to tell us that Percival Lowell was correct in thinking that Mars had canals that were built to carry water from the poles to fields of crops and that these canals provided water to inhabited cities."

Tolya thought this response was a bit odd and off the subject, but he continued in a moderated tone.

"No, only that life may have existed before. Even though we did not find it during mission, maybe it will be found in subsequent expeditions to Mars. How do you say in English: 'the absence of proof is not proof of absence.' "

The group became silent with this remark. Tolya had noted a few minutes ago that John had finished his check-out and had departed for the sleep pod area. He looked at his watch.

"I must leave now to do piloting simulations on my sleep pod PC. Good night."

He departed for the hatch leading to the upper level.

Mike finished his drink, then floated over to the mid deck telescope to take a gaze at the Earth, an activity he never tired of.

Jango stayed by himself in the dining area and reflected angrily at Mike's comments. He had suffered a loss of face, as if his abilities were being questioned in public.

I will show them, he thought.

This sense of having to prove himself and show others his abilities was not a new experience, for he had faced such challenges all his life.

Jango's actual name was Wang Jianguo. He had been born 40 years earlier on a wheat and rice farm not far from the Huai He River in north-central Anhui Province in China. Being the second son, he had little chance of taking over the farm, or anything else for that matter, despite the meaning of his given name: "build the country." Instead, he applied himself in school and did well enough to be accepted to Shanghai University. At first feeling inferior to many of the more sophisticated city-born students, a feeling contributed to by his slight build and shy personality, Jango worked compulsively hard and gradually began surpassing his classmates academically. Reflecting his childhood growing up in the country, his interests gravitated to the Earth sciences, and he subsequently earned a Ph.D. in applied geology and mineralogy.

Picked by the Chinese Space Agency as one of their astronauts to visit Lunar Base Alpha in 2027, he conducted analyses of the lunar rocks that were obtained near the base, writing several brilliant papers on his findings. When China was offered a chance to participate in the *MarsExplore* Expedition, Jango was one of two final candidates. His more sociable countryman was initially selected, but he was killed in a plane crash shortly thereafter. As backup, Jango was ultimately picked to represent his country. Along with Katya, he was specially trained to handle exobiological material and was tasked with finding evidence for life in rock and soil samples collected on the Martian surface. Aware that he would be away from his wife and son for several years, and that he would be the only Asian member of the crew, this introverted scholar had many hesitations about going. However, he could not refuse the honor to his family, despite feeling awkward and different from his fellow crewmembers, and he looked for ways to retreat to the quiet safety of the computer in his pod.

Part of his discomfort reflected political issues beyond his control. After the turn of the Century, China had begun to aggressively pursue a national space program, which culminated in the launching of its astronauts into space, the building of its own space station, and the landing of its manned rockets on the Moon. These activities stimulated space programs in the United States, Russia, Europe, and Japan to become more active, and a new space race developed. With the growth of space tourism and related businesses, as well as the scientific benefits resulting from the monitoring of global warming from orbiting satellites, governments and agency heads promoted the advantages of increased involvement in space, which included a strong manned presence. In addition, the military establishment in several western countries recognized important security benefits from Earth-orbiting satellites and bases on the Moon, especially with the growing Chinese presence. But what captivated the public was the idea of going to Mars and perhaps finding life on its surface.

Although a manned mission to Mars had taken place in the mid-2020s that was sponsored by a consortium of wealthy individuals and businesses, it merely circumnavigated the planet. It became apparent that a landing expedition would be too expensive for one or a few individuals, corporations, or even countries to support. The issue was debated in the United Nations, and it was decided that a broad-based consortium of government and business interests were needed that transcended national boundaries. The promise was made that new technologies would result that would help deal with the global warming crisis and increase corporate profits, as well as instill hope that humans could expand further into the heavens, thus relieving some of the people-caused environmental pressures on Mother Earth.

As a result, the World Space Council was formed to coordinate space activities and policies. The space agencies rushed to join, pooling their resources and vying for crewmember slots in a future *MarsExplore* Expedition. They were partnered by corporate and individual interests, and this mission became truly international and wide-ranging in support. China remained the lone holdout for a few years, but it finally joined as the crew was being selected and they were promised one crewmember slot as an inducement. Council members were happy to include China, partly for its financial and technical contributions, but also to absorb in the fold the last remaining competitive space program. However, some of the smaller members of the consortium were upset that their astronaut candidates would be excluded in favor of the Chinese newcomer.

Jango became acutely aware of this resentment during his training. The problem wasn't so much from his fellow crewmembers, who tried very hard to include him in all their social activities and customs. In fact, following the Russian tradition of diminutives, he had been given his nickname by Tolya (whose own name was Anatoly Polyakov), and it had stuck. Instead, he felt more pressure from the international media and the training personnel from his own country. After all, he wasn't China's first choice. But he was determined to represent his country well and to show the world that it had not made a mistake by including him on the mission. However, his self-consciousness and introverted personality was seen by many as unfriendly and guarded, and he remained an outlier in comparison to his more affable crewmates.

8 Tolya

The upper deck contained six sleep pods, three on each side of a central hall. Each was essentially a three-meter cube. The curved outer wall contained a small porthole looking out into space. The inner wall had a lockable door for privacy, which was given a high priority on such a long mission. Each pod contained a horizontally-oriented bed with a mattress that was attached to a frame with Velcro straps. This traditional construction was intended to preserve a sense of Earth-like comfort and security when going to sleep. Above was an overhead light, and to the side there was a chair with Velcro-tipped legs and a small desk with an attached lamp and personal computer. Straps were connected to the bed and chair to secure the crewmember from floating in the microgravity. There was space above for shelves and lockable cabinets to hold clothes, toiletries, and personal items. At the end of the central hall was the bathroom, containing two small private toilets, a washing area, and an enclosed shower.

Tolya floated up from the mid deck and emerged from the hatchway opening, his eyes locking on Juliette's pod, which was just across the hall from his. He became excited as he anticipated being with her. Juliette had never invited him to her sleep pod before. In fact, she had never really encouraged his advances until now. Perhaps she was becoming interested in having a relationship. He confessed to himself that he was especially interested in her.

Being isolated together for nearly two years is reason for more than collegial friendship, he thought.

His special feelings for her were a new experience for him. As a career pilot, he was trained to be cool and to deal with both emergencies and routine events with equanimity. He rarely felt strong emotions. At most, when he was under high stress, he might experience some abdominal discomfort. One time on a lunar mission, the oxygen generator broke down half way to the Moon, and there was some question if they would make it to the base before the oxygen ran out. Some of his fellow crewmembers were visibly nervous and bordered on panic, but he remained outwardly calm. However, he felt some discomfort just below his stomach. This caused him to remember how a burst appendix had killed his older brother when he was a boy, and he secretly began to worry that he now had appendicitis. After the crew managed to land on the Moon with the help of the reserve oxygen from their spacesuits, Tolya continued to obsess about his appendix until he went to the base doctor and was cleared. When during his examination he verbalized only casual concern about the broken generator, the doctor thought that he had been expressing his anxiety through bodily symptoms rather than through his feelings. Since then, he tried to be more aware of his emotional state. Although this was sometimes difficult for him, he was making progress. He wondered if his growing feelings for Juliette, not all of which were sexual, were an indication of this progress.

Everything was quiet on the upper deck. Tolya suspected that John was in his sleep pod and had settled in for the night. He floated over and quietly knocked on Juliette's door, and she opened it.

"Come in. I have the game ready to go," she said.

She was dressed only in a tee shirt that reached down to her upper thighs and was secured by a belt slung low on her hips. The hem wanted to float up in the microgravity, and she demurely held it down with one hand. Tolya found himself becoming aroused.

She ushered him in and sat on the bed, offering him the adjacent chair. They both strapped themselves down, and she strapped the chessboard to her uncrossed legs. As she placed pieces on the board, Tolya noticed two mini-squeeze bottles of Cognac Velcroed on the floor next to the table.

"Where did you get those?" he asked.

"I snuck them away from the party…in case I couldn't sleep," she added with a smile.

She began the game by moving the white Queen's pawn two spaces forward. The pieces were magnetically held to the chess board to prevent them from floating away.

"Are you happy about going home?" he asked to make conversation after moving his own Queen's pawn. His eyes kept jumping from the board to the rim of her exposed thighs.

"Yes, although being on Mars was truly wonderful. We proved to everyone that humans can live and work on another planet. And it really was a special place, with much to see. Having the Rover really helped."

"It was shame that we could not reach *Olympus Mons*," said Tolya. "It would be great to tell people in cosmonaut center I climbed largest volcano in solar system!"

Juliette looked up at him. He found himself becoming entranced by the deep brown of her eyes. She nonchalantly released the clip holding her hair, and it began to float wildly in the microgravity. He wondered what it would be like to run his hands through it as she looked down again at the chess board and resumed the conversation.

"I was happy to be on one of the Rover excursions to *Valles Marineris*. I am still amazed how large it was. When I was at Cal Tech, I took a trip to the Grand Canyon. It was like a gopher hole by comparison."

Tolya laughed at her comment. The two of them continued to play chess until she broke the silence.

"You know, Tolya, I heard about the missions you piloted to the lunar bases and the medal you received for saving those people who crash-landed near Lunar Base Beta, but I don't know how you got interested in being a cosmonaut in the first place."

"My father was in Russian Air Force and was assigned to many places. He received Hero of Russia Award. He took me on training flights when I was boy, and I got interested in flying. I joined air force and later became test pilot."

"Then how did you get to be a cosmonaut?"

"My father always spoke about glorious Soviet and Russian space program. I decided to apply to be cosmonaut and was selected. I trained at Gagarin

Cosmonaut Center, finishing in 2020, then was assigned to lunar program, then Mars, and here I am."

"You sound like you have had a busy life. Did you ever think about getting married?"

He answered the question without showing the surprise he felt at its abruptness.

"I had many relationships with women, but I was sent to many places, and I did not want to have mobile family life like when I was boy. Plus, I could have blown up test plane or crash-landed space vehicle, so it was better not to have family while I was pilot. Maybe after we return to Earth, I will think about it. Since we all are being exposed to lifetime dose of radiation during this mission, none of us will ever fly again, so that would be time to settle down."

"Yes, the radiation exposure certainly will change our lives. I worry about not being able to have children, although the flight surgeons believe that our cumulative dose will be under the critical amount to affect fertility."

After he did not respond, she continued.

"How exactly did you get your medal?"

"There was supply ship going to Lunar Base Beta. They had problem with engine and crashed. Somehow, crewmembers survived. Me and my copilot were sent to rescue them in taxi ship. There were many rocks in area, so we could not land at crash site. Instead, we landed five kilometers away, activated wheels of taxi for land travel, and drove to them. We found them, just barely alive. To me, it was part of job, but because one of crewmembers was from Japanese Space Agency, it made international news. So, I got medal from World Space Council."

"You should be proud of yourself."

"My father was very proud. He always joked to his Air Force buddies that we were two-medal family."

"Check," she said.

"What?" he responded. Then, looking down at the board, he declared: "And mate. You got me."

He glanced at Juliette, saw that she was smiling in a friendly but not gloating way, and was immediately captivated by her.

She is real beauty, he thought.

He leaned across the board, and when she did not move back, he gave her a light kiss. She enthusiastically returned it. Loosening his microgravity strap, he floated over to her, held on to the rim of the bed with one hand, embraced her with his other arm, and kissed her again. He could feel her breast against his chest, and he became further aroused. He became aware that she had removed the chessboard from her lap, and as it floated away the hem of her shirt rose up. In one smooth and fluid motion, he gently maneuvered her onto her back. He loosely strapped them in place, all the while embracing and kissing her as she pulled his shorts down his legs. Grabbing both sides of the bed, he alternatively pulled the two of them down onto the mattress, then let them float up against their loose tethers. As they became increasingly aroused, they both reflected with joy on the wonder and uniqueness of this experience.

9 Interlude

It contemplated recent events and realized that the changes It had made were generally successful. The alterations to the computer program that blocked the thruster pressure warning system were implemented. The computer alarms were not triggered when It caused the thruster fuel pressure to drop. Fortunately for the bipeds, a routine maintenance procedure had alerted them to these changes. However, since many of the routine maintenance checks since launch had now been completed, the chances of the bipeds making future discoveries were lessened. There was no need to change the course of action.

But the bipeds had been warned. This did not mean that the Plan was wrong, only that the next step should not be implemented too soon. With the passage of time, the bipeds would likely become less alert, less vigilant. There was no need for haste.

10 Katya

The morning after the party, Ekaterina Malenkova floated down to the dining area with great anticipation. The rustling noises coming from Juliette's room last night kept her awake for a while, and Katya wondered if an intimate French-Russian connection had been made at last! The other crewmembers had whispered about the potential that existed between Juliette and Tolya, yet the French woman had been strangely cool to the handsome pilot. Maybe she preferred women? But Katya had seen her medical file, and there was nothing in her social history to suggest this. Maybe she had her sights set on someone else in the crew? Certainly, Mike had been friendly toward her in the past, but Juliette didn't seem interested in him as a partner.

No wonder, Katya thought. *The American engineer has been intolerable of late. He has been overly critical and intrusive. Some of his comments are off the wall. And his clothes! Bright yellow tee shirts don't go well with red shorts.*

Emerging through the hatchway opening, Katya glanced around the mid deck. John and Jango were finishing their breakfast. Mike had already eaten and was using his free time to look at the Earth through the telescope. But no Juliette or Tolya. Were her suspicions warranted? She playfully contemplated going back upstairs and knocking on Juliette's and Tolya's doors for some made-up reason to see who was where, but she decided that it would be better to see how things proceeded naturally. Besides, how would it look for the trusted crew physician to assume the role of "outing" a younger man and woman experiencing the joys of love in microgravity?

In fact, she was very aware of her role in fostering cohesion in the crew. Given her scientific background, her years of clinical experience, and her knowledge of medical and psychological issues involving space travelers, she was a good choice for the mission. In addition to these professional factors, having a second woman on board made the crew more gender-balanced. She shuddered at the thought of having a crew of five men and one woman. The lone female crewmember likely would feel psychologically isolated and always in the spotlight, and she could be scapegoated by the men when things went wrong. She appreciated the fact that sometimes she and Juliette gave each other female support and occasionally could have some "girl talk." Along these lines, she had always thought that Jango's being the only Asian in the crew created stress for him and probably resulted in his feeling culturally isolated, although she had to admit that his introverted personality no doubt contributed to this as well. Nevertheless, she felt sorry for him and did what she could to involve him in group activities.

Katya had always thought that John was a good choice as Commander. With slightly graying thick hair and steely green eyes, he looked the part of a can-do leader. Over the years she had known him, he had acquired an excellent reputation for commanding space missions, but he was also recognized as being psychologically-minded and relating well to people. And being the only two 50-year olds on board, she felt that the two of them supported each other and could be seen as mature role models by the younger crewmembers.

After all, John and I are the mother and father of this family! she thought with a smirk on her face.

But this notion gave her a pang of sadness. It was thoughts of family life that made her think of Slava and his painful death from cancer five years ago. In fact, she likely would not be here but for his influence.

As the daughter of two physicians in St. Petersburg, Katya had always been interested in the life sciences. She had studied microbiology for a while in college, then decided to go to medical school, where she met Slava. The son of a pilot, he was passionately interested in aerospace medicine, and after they were married the two of them began working at the prestigious Institute for Biomedical Problems in Moscow. Both became flight surgeons and worked with cosmonauts, although Katya continued her research interests as well by studying the effects of microgravity on bacterial growth in space. The couple decided not to have children, given their high work demands. Katya sometimes had regrets about this decision, but the two of them were devoted to each other and to their careers, and their lives were fulfilling. After Slava died, Katya had no commitments on Earth, so applying for the *MarsExplore* Expedition seemed like a good idea. Her only living family member was a sister in St. Petersburg, but they rarely saw each other due to her work duties in Moscow.

There was a great deal of competition for the position of *MarsExplore* physician. However, her medical and research experience and her complementarity to Juliette all contributed to her being selected. In addition, she had spent time in Houston Mission Control at the Johnson Space Center supporting the cosmonauts involved with some of the lunar base missions, where she became fluent in English and comfortable with the NASA space culture. She worked with John at JSC as well, which improved her chances for selection after he was picked to be Commander. As a spiritual person, she was sure that somewhere Slava knew about her selection and was proud of her.

"Good morning," she said as she floated over to the food dispenser.

"Good morning, Katya," John responded. "How did you sleep?"

Suppressing a smile at being kept awake by the revelry in the sleep pod next door to her own, she said: "Very well."

"So, are you going to give us a favorable report on our exercise regimen?"

Knowing that the required daily two-hour exercise program was not strictly adhered to by all of the crewmembers, she diplomatically responded: "All seems to be going well. Everyone's vital signs are stable and within normal limits, the tests indicate minimal calcium loss, and muscle mass looks good, according to the leg circumference measurements."

"Great," John continued. "I just received an e-mail from Mission Control. Everything about our return flight looks good from their end. They also wanted to make sure that the crewmembers are healthy and happy, and they asked for a formal report about crew health. Could you please send them something?"

"Will do, John."

"They're always wanting something, always snooping around," Mike commented from his seat at the telescope. "Mission Control doesn't do squat for us. I don't know why we bother responding so much to their demands."

In a way, Mike is right, Katya thought. *Given our long distance from home and the communication lag related to this distance, we are quite independent and autonomous from Mission Control. But maybe Mike is displacing some tension he is feeling internally or with another crewmember.*

This reminded her of the group dynamic studies that were reported around the turn of the Century which suggested that some crewmembers in space transferred unpleasant emotions they were feeling toward each other to safer and more remote individuals located in Mission Control. Although helpful in the short term, in the long term such displacement did not correct intra-crew strife but allowed it to fester over time.

I better watch him more carefully, she thought further. *His irritability seems to be getting worse and may become disruptive.*

John ignored Mike's outburst. "They also sent us some news reports concerning our mission. They mostly were positive, but here and there were comments about our failure to find evidence for life on Mars. 'Just a few red rocks!' one of the headlines read. Maybe we should have taken some toy dinosaurs and alien dolls along with us to plant in the soil and then 'discover'."

Everyone laughed, although most were sympathetic with his sentiment.

"I guess since we were on Mars, we could call ourselves Martians, maybe the first who ever lived there," said Katya. "Of course, according to the panspermia hypothesis, we were just going home."

"I don't believe in panspermia," grumbled Mike. "I just can't buy the idea that life on Earth was seeded by microorganisms from space."

"Then what about those rocks we were shown during our pre-launch briefings? You know, the ones that were found on Earth but were actually ejected from Mars as a result of meteorites crashing on its surface. Some of them contained impurities that looked like bacteria fossils."

Mike rolled his eyes. "But we also were told that this idea has never been proven conclusively. Plus, how could microorganisms survive the stress of a meteorite impact and the trip through space?"

"Microorganisms are hardy little creatures." Katya responded. "And rocks are good insulators. We know that living bacterial spores have been able to survive in Earth-orbiting satellites for several years. Plus, extremophilic microorganisms on Earth are able to live in a number of hostile environments, such as under the Antarctic ice, around volcanic vents and hot springs, and attached to rocks."

She thought further.

"And don't forget those studies on Earth where methane-producing organisms were able to survive in sealed chambers that contained Mars-like soil and atmospheric conditions."

"Yes, but in those studies they needed carbon dioxide and hydrogen from water to live. There is plenty of carbon dioxide in Mars' atmosphere, but we sure didn't find any rivers or lakes on the surface!"

Mike chuckled at his own comments, then turned to Jango.

"But I will admit that we found high levels of methane gas coming from that cavern we discovered during our expedition to *Tharsis*. Remember, Jango? I wonder if..."

"I told you at the party yesterday that there was no evidence for life in there," he interrupted coolly.

This brought silence to the room.

Mike then turned to Katya. "How about you? Did you ever find anything interesting from the samples you collected close to the base?"

"No. Nothing grew out on the culture media, even the samples that were incubated in Mars-like soil and atmosphere."

"Good morning everyone," a smiling Juliette announced as she floated out of the hatchway opening. "It sounds like you are all having a spirited conversation."

She certainly has a curious look to her, thought Katya, *glowing yet slightly embarrassed, I would say.*

"We were just discussing how well the little green men on Mars were able to hide from us," joked John.

"Yes, I guess they were," she responded quizzically.

John changed the subject. "Juliette, today you and I should program in some of my navigation figures to make sure that they are compatible with those generated by CARS. By the way, did you see Tolya before you came down? We need to include him as well."

"Here I am, Commander," the pilot responded as he emerged from the hatchway opening. "Good morning to everyone on this fine day."

He is trying to suppress a smile, Katya thought of her fellow countryman. *He looks more guilty than victorious, like a little boy caught sampling the borscht before a meal. But the bravado is gone. What did she do to our rakish pilot?*

"Tolya, I need to go over some navigation figures with you before we try them out on CARS. We need to make sure that our brainy computer is still thinking clearly."

"Any errors are likely human ones," said Tolya. "CARS never makes mistakes, only humans. I know it is difficult to accept, especially since he controls so much of mission and life support, but it is true!"

John considered the choice of male pronoun for CARS. In a sense, it was hard not to think of the computer as human and a member of the crew. It certainly played a role in most of their activities and had managed to keep them alive and functioning so far.

John responded: "It has indeed been a marvelous aid on board, one that has freed us up from a number of tasks, both the mundane and those that are beyond our abilities. But we still need to do routine maintenance checks and monitor its performance. Especially since the yaw pressure episode. Juliette, do you agree?"

She nodded.

"OK, so after you two finish breakfast, we can check out our computerized 'peripheral brain' to make sure that it's doing its job. In the process, we can reassure ourselves that all is well." He concluded, dryly: "I guess this would be simpler and more direct than having a weekly psychotherapy session."

11 Mike

As his crewmates went back to their breakfasts, Mike Lipinski continued looking at the heavens through the telescope. It calmed him down and helped relieve some of his growing agitation. He never tired of seeing the Earth. It looked like a fragile blue and white jewel through the eyepiece. In this activity, he was not alone. Since the beginning of space travel, astronauts had been captivated by observing their home planet floating in the heavens. Nearly all saw this experience as the major factor contributing to their having a positive reaction to being in space. Some were struck by the lack of obvious national boundaries and the consequent unity of mankind. Others took a spiritual perspective and perceived the Earth as God's wondrous creation. Still others enjoyed the aesthetic beauty of what they saw, with the striking blues and whites and browns and greens, all intermingling, as if they were pigments on an artist's palette.

Reflecting on John's comment about checking out CARS, Mike thought: *This is my psychotherapy.*

There was one other emotion that he was experiencing. The view of the Earth made him feel special. He could alter the view, make it larger or smaller, move it at will. It was as if he was the Master of the Universe, completely in control of his home planet and all that it represented. He felt a sense of power.

Mike had not always felt this way. He never knew his real parents, spending his lonely early childhood in a Catholic orphanage in Chicago. At the age of six, he was adopted by a Polish family who had links to the orphanage through their church. His father drove a delivery truck, and his mother worked part-time in a department store. But Mike proved to be a brilliant student, scoring several grades above his classmates on national tests and effortlessly making straight A's in his classes. He attended Northwestern University on a full scholarship, where he double-majored in mechanical and electrical engineering; the California Institute of Technology, where he received his M.S. in Aeronautical Engineering; and the Massachusetts Institute of Technology, where he received his Ph.D. in Aeronautics and Astronautics.

Although moody at times and of unremarkable appearance, his intellectual abilities nevertheless made him appealing to many women. During graduate school, he briefly married but then divorced one of his classmates prior to receiving his doctorate, deciding that the work involved in sustaining a relationship interfered too much with his destiny in life, which was to be at the forefront of the engineering field. After being accepted into the astronaut corps, he flew on three short lunar base missions as the flight engineer, and then he was accepted for the *MarsExplore* Expedition.

And here I am, he thought. *The chief engineer of the mission, the one who knows the most about the workings of the ERV. If anything goes wrong, I'll fix it.*

Gazing over at Juliette for a moment as she finished her breakfast, he thought: *I wouldn't mind having sex with her, though, if only we could get Tolya out of the way. I can't see how she would ever be interested in him—he's all show and not nearly as smart as I am. Up until now, she has not seemed interested in anyone. But it's curious that she and Tolya both came late to breakfast today…*

John interrupted his reverie: "Mike, how does Mother Earth look today?"

"Just beautiful, John. Getting larger and larger by the week. We are on our way home."

"Yes we are. I am certainly looking forward to it. How about the rest of you?" John asked, making conversation but also trying to get a sense of the mood of his crew.

"I just wish we could get home sooner," volunteered Katya. "More than half a year to go. One can get a bit impatient thinking about how long we still have."

After a brief silence, Juliette volunteered: "It is a bit long until we reach home, but I am finding things to do to occupy my time."

Tolya tried unsuccessfully to suppress a brief smile.

"I am having no problems," said Jango. "My programming activities and computer games are helping me fill my time."

"I am tired of practice landing simulations," said Tolya. "It is difficult to think about landing when we are so far away from Earth. At least simulator has built-in emergency scenarios to keep me entertained. It is like winning at chess."

It was Juliette's turn to smile.

"Katya, you seem to be keeping yourself busy," said John. "I guess making sure that we are all in good health continues to be a full-time job."

"Yes, it does." She looked around the breakfast area. "All of you have specialized functions to keep you active at certain times but not at others. John and Tolya are busy during take-offs and landings. Juliette has her periodic CARS checklists to complete. Jango and Mike had a lot to do on the Martian surface, with Jango collecting and analyzing soil and rock samples, and Mike monitoring the systems in the habitat and Rover. Now, of course, Mike has his periodic servicing of the ERV. But whether or not each of you is busy, you need to function at top physical and mental shape throughout the mission, and that is my job. So I am always busy. But it is a pleasant challenge."

12 Irritability

Later that afternoon, Juliette was running an analysis of CARS and its operation from her sleep pod PC. Try as she might, she could not find any indication that CARS had recorded any kind of recent pressure malfunction in yaw thruster #1. Yet according to Mike, the local gauge had shown a pressure deficit. It was all very strange.

Just then, she looked out her open pod door and saw the chief engineer floating out from the bathroom.

"Mike, could you come here a moment?" she asked.

He came over and positioned himself at the doorway.

"What do you want?" he asked tersely.

"I cannot find any CARS evidence for an anomaly in the yaw pressure system yesterday, or even for the past week. This is very confusing to me. Are you sure you found an anomaly? I can't confirm it."

Mike reddened.

"Of course," he said. "I am a top engineer—I know how to read a gauge. The fuel pressure in yaw thruster #1 was low, and I corrected it. Period."

"Well, I was just wondering…"

"And don't try to blame me. CARS messed up, and you are responsible for its operation."

He floated in right up to her desk.

"You know, you never give me enough credit. I am the chief engineer here. I was tops in my class, and I have served on many important missions before this one. And you? You are a computer jock. Or rather, a computer jockette… with a jock strap!"

He snickered at his comment, then continued.

"And you have eyes for that Russian, that phony Casanova. He is all show and no go. I can walk around him any day when it comes to fixing things."

He put his hand on her shoulder and began to massage it. She recoiled, shocked at his anger and informality. Although he had frequently been irritable and cocky during the mission, it had never reached this degree of unpleasantness.

"I do not know what has gotten into you," she said, "but I want you out of here. You are bothering me."

"You called me over," he responded. "You don't know what you want. You should feel lucky that I am paying you any attention. I can help you fix that damned computer. But I will expect you to be nicer to me, to give me proper attention and respect."

He moved closer, maneuvering his face towards hers. She quickly responded by pushing him in the chest. Because she was strapped down and he wasn't, he floated backward, partly out of the room.

"What's going on in there?"

It was Katya, who was just emerging from the hatchway leading to the mid deck.

"Nothing," Mike responded. "I was just helping Juliette fix CARS. But we're done." He glanced menacingly at the French woman. "Good luck."

He floated by Katya and went down the hatchway.

The Russian physician entered the room and closed the pod door behind her.

"Are you OK? What happened?"

Trembling slightly, Juliette said: "I don't know what has gotten into Mike. I called him over because I couldn't find a problem with CARS, and I wanted to discuss what he had found with the pressure system. He became angry and threatening, took it personally. At the same time, he came very close to me. I thought he was going to hit me…or kiss me…I don't what he was going to do, but he was very intrusive!"

"Did you provoke him in any way?"

"No, I just wanted information. He was insulting."

Tears began to well up in her eyes. Katya put her arm around Juliette to console her.

After a moment, she became calm and said: "I am all right, Katya. Maybe I said something to irritate Mike. I don't know. Maybe he will cool off. Nothing bad happened."

"Fortunately, Juliette. But from what I saw of his behavior, it was not acceptable. I will discuss this with John. In the meantime, you should give Mike some distance until we can figure out what is going on. OK?

"OK, Katya. And thank you."

As the physician left the pod, she was troubled. Mike's behavior was not normal, even for him. He seemed pressured, irritable, almost irrational. Maybe it was the stress of the mission getting to him, or maybe it was something worse. She would need to monitor him very carefully.

13 Sabotage

"Have you found anything regarding the yaw pressure issue?" asked John, floating over to Juliette's pod just before dinner.

"CARS recorded no history of an anomaly in any of the thrusters," she responded. "But taking Mike's report into account, I made a step-by-step analysis of the thruster programs. I just found something odd. There was a specific instruction added in the programming telling CARS to ignore any fuel pressure change in yaw thruster #1. So even if there was a drop in pressure, CARS would not record it or trigger an alarm."

"How could something like that happen? I thought the programs were set before the mission."

"They were. The change occurred after the ERV was launched from Earth, probably even after we left Mars, since the thrusters behaved just fine during the deorbiting burn."

"Should we be alarmed by this? Is there a chance that one of the life support systems is similarly defective?"

"I checked the major systems, and everything else looks fine. Not only are the gauges and alarms functional, but all of the life support programs seem to be intact."

John thought for a moment, then asked: "What do you think is going on?"

Juliette furrowed her brow.

"I don't know of any natural process that would have affected the programming for this thruster in isolation from the others. If there was a power surge or a solar flare impact event, other systems would have been affected as well, and we would have been aware of this. In terms of a crewmember adding the new instructions, we all have access to CARS through the command and laboratory terminals or through our sleep pod PCs. However, it couldn't have been an accident."

"What do you mean?"

"Like other vital systems, the thruster programs have coded firewalls that block access, and I don't see how anyone could have broken through by mistake or by routine computer usage."

"So it would have been a conscious effort. If someone wanted to break the code, would that have been possible?"

"Yes. All of us are pretty computer savvy. It would take a very persistent attempt by someone really motivated to want to change the thruster feedback loop, but it would have been within the capability of all of us."

The two of them were silent until John broke the ice. "What can we do to deal with this?"

"I went ahead and deleted the erroneous instructions in the thruster detection system," Juliette responded, "so everything is nominal again. I checked all of the other thrusters and the main engine, and everything looks fine."

"What if another system becomes flawed in the same way?"

"Well, we have a good maintenance program where all of our systems are checked periodically to verify that CARS is working properly. After all, Mike discovered the thruster problem during a routine check of the engine pressure systems. I suggest we continue to monitor things as scheduled, but be very alert and vigilant to possible programming changes."

"That makes sense. Finish up here, and then we'll input the navigation figures."

As John floated away, he was troubled. Her words echoed in his mind: "It would take a very persistent attempt by someone really motivated to want to change the thruster feedback loop." What if that were the case? What if one of the crewmembers was cracking under the psychological stressors of the trip home? What if someone wanted to sabotage the mission in some way?

He went down to the mid deck and found Katya alone in the lab.

"I just spoke with Juliette about the thruster pressure episode," he said, floating over to her.

"Did she find anything?"

"There was an alteration in the CARS program that she thinks may have occurred after we left Mars. It would have prevented us from knowing about problems with the thruster pressures."

"Does she think that other programs were affected as well?"

"I asked her that, but she thinks not. She advises that we be especially vigilant when we do our systems maintenance checks. I guess that would apply to the medical and lab systems as well."

"OK John, I will keep tabs on my section."

"And Katya, maybe you should include a complete psychological evaluation in your future routine medical examinations. Quietly of course—don't make a big deal about it."

"Do you think that someone is having emotional problems and that this is related to the thruster issue?"

"I don't know. I just think that we need to be vigilant. I suppose we all are a bit homesick and anxious for the mission to end. Under these conditions, someone could get unhinged and act irrationally. Or out of boredom, someone could play with his or her personal computer and cause a program change to CARS. Who knows?"

"But isn't there a firewall that protects…"

"Yes. But maybe this breach was not a mistake."

After a momentary hesitation, she asked: "Do you think someone consciously made the changes? Perhaps to sabotage the mission?"

"At this point, there is no evidence for such a thing. I just don't know who would do that, or what the motive would be. Anyway, I would appreciate your helping me monitor the mental health of the crewmembers, including you and me."

Katya became quiet, then said: "John, I need to tell you something. I know there is physician confidentiality on board, but an incident occurred earlier today that may be related to what you just said. And as Commander you have a need to know. Earlier today I saw Mike encounter Juliet in a very hostile and intrusive manner. From what I saw, he overreacted to her questioning him about what he found when he discovered the yaw pressure anomaly."

"What! I just saw Juliette. She didn't say anything to me. How is she?"

"She is doing OK, but it is not surprising that she didn't tell you. She is probably embarrassed by the incident and maybe thinks she contributed to it. He made menacing and perhaps sexual advances toward her. When I saw him, he seemed about ready to lose control. And given his status as the chief engineer on board, that would be a bad thing."

"Damn right, it would be a VERY bad thing. What do you think we should do?"

"I think we should carefully watch his behavior and mental state. If another such incident occurs again, we may need to formally intervene, perhaps with medications or restraints."

"That wouldn't be easy here—we really don't have the proper facilities to restrain anyone."

"I know, but there are straps on the exercise table. And I will keep the medications handy just in case."

"OK. But what about the other crewmembers? Do you see any evidence of anyone else acting crazy?"

"No, not really. Jango continues to keep to himself a lot, but not more so than usual. Everyone else seems to be about the same as when we left Mars."

"Good. Let's keep each other posted. Especially about Mike."

As he floated away, Katya thought to herself that what John was struggling with was indeed alarming. By giving the crewmembers the ability to communicate with CARS through their pod PCs, the mission planners had thought that it would maintain morale by allowing everyone to access a variety of manuals and systems related to their work, along with leisure-time books, movies, and computer games, all in the privacy of their pods. No one had thought about what would happen if this system were misused by someone who was unstable or had malicious intent.

I certainly will be vigilant, she promised herself.

14 E-mails

To: swood@hotmail.com; mwood@hotmail.com; rwood@utexas.edu
(RESTRICTED USE)

From: john.c.wood@MarsExplore.gov
Subject: Hello
Date: Sunday, 6 May 2035, 1520 hours CDT

Hi Everyone,

It has been a while since I wrote to you. All is going well with our mission. We had a nice May Day celebration, and we are preparing for our mid-course maneuver in July, which will fine-tune our trajectory home.

We have set up a return schedule that works pretty well for us. It is similar to the schedule I sent you earlier when we were going to Mars. We get up around 0730 hours (for the return, our time has been synchronized with JSC). After dressing and checking our PCs for messages from Earth, we meet and have breakfast together to work out the day's events. Our first work period starts around 0900. Some of us exercise on the treadmill and exercise table or do vehicle maintenance, while others study the Martian rocks or write papers from our studies on the surface. We eat from 1200 to 1300, then start our second work period. Those who haven't exercised do it then. At 1600, we have our daily conference, where we review the day's activities, plan for the next day, and talk about any problems on board. We eat dinner from 1700 to 1800. Then we have free time to catch up on anything we didn't get to earlier, read or watch a video, or whatever. Last night, some of us watched the classic movie "2001" and wondered who was more bored, the crew in the movie or us in our ERV! It was great until the HAL computer went berserk in the movie, then we all got a little worried about CARS. Ha! Ha! Bedtime is usually at 2200, with the cabin lights dimming at 2300. So, that is our typical day.

As I said in my last e-mail, we have a lot of rocks to bring back from Mars. We continue to look them over, but we still haven't found any evidence for life. This is disappointing. We have had several discussions lately about what we did find, and Jango maintains that the geologists on Earth will be happy. The main thing is that we showed that people can live on the surface of another planet for more than a year and survive. With more expeditions being planned for the next several years, we will continue to learn and maybe one day establish a permanent settlement. You kids may someday live on Mars as citizens!

Stephanie, how is your mother doing? The flu can be very difficult, especially for someone her age. Say hello to her and your father. I hope you are doing well, too. I really miss you. This is the tough time in the mission, when all the excitement is behind us, and we are just waiting for the return trip to end. It's like going on a vacation to Europe, when you are busy reading about what you are going to see on the trip over, then enjoy the time while you're there, but the return home can seem to last forever. Everyone here tries to support one another—we are all in the same boat. How did everything work out with the bank loan? Let me know if I can help in any way.

Melanie, how are your prom plans going? Are you going with Jim? Are you excited about your graduation? Darlin', I'm sorry not to be there in person to see you get your diploma. It's exciting to see you reach this milestone. I know you'll enjoy yourself at UT. Austin is a great place.

Robert, congratulations on your recent e-mail about making the Dean's List last semester! It sounds like you're off to a good start again this semester. Have you decided yet on graduate school? Your summer research job at JSC sounds pretty exciting. When we land, it will be Thanksgiving, but I know you will be off again at college. A senior! Pretty cool.

If for some reason all of you can't be there for our landing, I will look forward to seeing you at home during Christmas.

I miss all of you. Take care.

Love,
Dad

****

To: john.c.wood@MarsExplore.gov (RESTRICTED USE)
From: mwood@hotmail.com
Subject: Hi Daddy
Date: Sunday, 6 May 2035, 2002 hours CDT

Hello Daddy,

It was great to hear from you! I miss you sooo much!! We can't wait for you to come home. I will definitely be at the landing, even if I have to miss my classes.

Jim asked me to the prom, and I am so excited. We will be having a space theme, and it will be so rad. I am on the decoration committee, and we plan to show a big display rocket coming back from Mars, just to honor your mission. Mom and I went shopping for my dress last week, and it is really cool.

It is blue taffeta, short, with a bow. I really like it. We will take pictures to show you.

I am also excited about my graduation. The plans are a little unclear, but it will be outside in the school courtyard. I hope we don't have July-like Houston humidity! The robes will be hot. Maybe I will wear shorts underneath (no one will know). I will get honors on my diploma, so you can be proud of me!

I am excited about UT in the fall. Robert has told me all the rad things about the place, where to eat, where to shop, who the best freshman teachers are…everything. Mom and I are driving up this summer to look things over again. I know she is a little sad that I will be leaving, but at least you will be back shortly after I go.

Mom has been trying to take care of everything since you left, plus work part time, plus be a mom. Sometimes it is hard for her, and she cries. But she has the JSC support group to go to. And she has been taking a night class in art at the community college. I think she really likes her class. Sometimes she doesn't get home until after midnight.

Well, don't worry, we are fine. I can hardly wait till you come back.

XXXOOO, Mel

****

To: john.c.wood@MarsExplore.gov (RESTRICTED USE)
From: swood@hotmail.com
Subject: Hello Dear
Date: Sunday, 6 May 2035, 2044 hours CDT

Hello John,

You sound well. The kids are fine, too. Mel is excited (as usual) about her prom, and we have picked out a nice dress for her. She is going with Jim. He is a nice boy, and I think things will be all right. Robert continues to do well at school, although I haven't heard from him in weeks. I guess he is moving along in life, as is Mel.

Mother is better. I have been going to see her in the hospital every day, and it has been a little stressful. Her doctor told me that she is very forgetful. In fact, she may have missed taking some medicines at home, since she came in to the hospital with high blood pressure. Dad is forgetful too, so between the two of them, I don't know how they can remember to do anything. We may have to talk about putting them in an assisted living program. Dad will be very resistive to this, but it may be time.

The banker wants to wait until you get back to approve the loan, since he read in the paper that this might be your last mission, and he didn't know what your plans would be as regards to staying at NASA. I tried to tell him that you would continue working in the space program, but he wondered what you could do to top being on the first expedition to Mars. He said he has dealt with many astronauts before who get frustrated by not being eligible for a future mission, and they end up leaving. Sometimes impulsively. Anyway, I think we can cover Mel's fees and living expenses for the first quarter from our savings, so hopefully we will get the loan as soon as you speak with him after you return.

I have generally done well, although I am a little stressed out with everything going on. It seems like you have been away forever. At my spouse support group, we were talking about all the pressures families feel when their husband/dad or wife/mom is gone for a long time. They all looked at me, since I am setting the longevity record for spousal separation. Don't get me wrong—everyone is proud of you and what you are doing, but it will be good for our family to reunite.

I have been taking an art class and really enjoy it. The teacher is very good. Most of the students are young, but one is about our age. He knows about your mission and how long you have been away, and he has been very supportive.

I hope everything continues to go smoothly and safely. It sounds like everything is pretty routine and without stress. I am glad for that—I was always a little worried when you were on Mars that someone would get hurt or get caught in a dust storm.

Say hello to everyone there, especially Katya.

Love,
Steph

****

To: john.c.wood@MarsExplore.gov (RESTRICTED USE)
From: rwood@utexas.edu
Subject: Yo
Date: Monday, 7 May 2035, 0040 hours CDT

Hi Dad,

Nice to hear from you. All is well here. I have been writing a term paper for my history class that is due tomorrow, and I just finished it. I wrote about the role of the space program in modern society. It focused on the societal impact of the major milestones, such as the first man in space, the first woman in space, the first person on the Moon, and of course the first crew to land on Mars. The topic has been fun to write about because I have lived part of it, even though I have been a little stressed out with all the papers being due the same week. I wish the profs here would coordinate things better.

I plan to take my Graduate Management Admission Test this fall, then apply to business school. I have been looking over school requirements and applications, and this has added to the things for me to do. I will give you a progress report when I see you after you get back. I hope to get away from school for the landing, but if not, then Christmas for sure.

I still plan to do the job at JSC this summer. It will be fun working there again, even though it will not be related to my graduate plans.

Have you heard about the Astros? They are nearly in first place, just one game out. I know it is still early, but I like the Astros' chances this year, especially with all those great pitchers. They have come a long way since last season.

Longhorns football also looks pretty good. I think they will be a Top Ten team this fall. Unless they choke like last year.

Well, that's all for now. I am starting to fade and will hit the sack—I have an early class tomorrow (8:00 AM—groan!).

Robert

15 Interlude

It decided that It should not kill them all. One of the bipeds would be useful to help It explain what happened and to be a back-up carrier for the progeny. But the rest needed to die, unfortunately but necessarily, or they might interfere with the Plan.

How best to do this? It realized that they were vulnerable to many things. They needed to eat and drink. Could something be put in their food or beverages? That might work, but it was possible that one or two would be busy and so not be eating with the others. Also, those getting sick might be alerted and take steps to heal themselves before they died.

It next considered that they needed to breathe. What about poisoning the air? If they breathed in carbon dioxide instead of oxygen, they would lose consciousness. The one that It selected as the companion could be revived before all were asphyxiated.

How to make this happen? It would have to gain control of the life support system, then release the right sort of toxic gas mixture. To do this, the feedback systems to prevent this occurrence had to be disabled, and the warning alarms had to be bypassed. Could this be done? It had done this once with the propulsion system, but perhaps further test trials would be useful.

16 Asthenization

As the days passed, Katya was aware that the long return home was continuing to take its toll. Although no further problems in the CARS system had occurred, the same could not be said for some of the crewmembers. As she was cleaning up in the lab one day, she reflected on the results of the psychological assessments that John had requested. Irritability seemed to be increasing among the crewmembers, especially with Mike. Jango was isolating himself more than usual, with the excuse that he was writing reports and working on his computer. Tolya and Juliette were spending more time together. Were they truly developing a relationship, or were they also isolating themselves from the others? John was trying to keep up morale, both his own and the crew's, but even he was more distant. He seemed to be preoccupied with his family back home, worrying about the effects of his long separation from them. Katya herself was aware of feeling fatigued, even burned out, impatient for the mission to end and spending more time alone in the laboratory. She told everyone that she was still examining the samples for signs of life, but she admitted to herself that this was an excuse for her to be alone and think. Like now.

I wonder if we are undergoing asthenization? she thought.

This syndrome was related to a neurotic condition, neurasthenia, which was first described in the United States in the mid-1800s. Katya reflected on its classic symptoms: excessive tiredness and fatigue, the two major complaints, but also morbid fears, hopelessness, irritability, concentration difficulties, and a variety of physical problems. When normally healthy cosmonauts who were flying in space began to report some of the symptoms of this disease, it was felt that they were undergoing a mild form of neurasthenia as an adjustment reaction to the isolation, confinement, and dangers of manned space travel. Russian flight surgeons called this reaction "asthenization."

As far back as the 1980s, Russian flight surgeons developed countermeasures to deal with asthenization. They found that the symptoms went away when crewmember morale was enhanced, such as with increased real-time communications with family, friends, and celebrities on Earth, or by sending up surprise presents and favorite food in resupply vehicles. Although such methods were effective during on-orbit missions, at the distance of the Mars crew from home, such strategies were not possible. The communication time delay interfered with real-time contact with Earth, and the ERV was too far away for resupply. The crewmembers were left to their own devices to deal with asthenization or any other problems that they experienced during the mission.

Then Katya had an idea.

Maybe the American July 4th holiday in two days can be used as a way of stimulating ourselves out of this...what do the Americans call it...''funk'', Katya thought. *We should plan more time for the holiday and make it as much fun as possible. I must talk with John about this.*

As she finished her lab clean up and left to look for John in the upper deck, she reflected further on the situation with Mike. Although he hadn't had an outburst similar in intensity to the episode with Juliette in May, he had continued to be irritable and intrusive when he was around other people. He also had been spending more and more time in his sleep pod, allegedly working on an important project that he would not clearly define. She wondered what could be happening with him. After all, potential *MarsExplore* crewmembers had been carefully screened for histories of psychiatric and substance abuse disorders before they were selected. They were also selected to be part of a cohesive crew in terms of compatible interests, respect for others, ability to work alone on a project when necessary, and possessing enough social skills to be able to interact with their colleagues at mealtimes and other social events. Although every person had a breaking point if stressed enough, there was no reason to suspect that any of the *MarsExplore* crewmembers would develop a major psychiatric problem like manic-depression or schizophrenia. So what was going on with Mike?

Perhaps he is simply having difficulty adjusting to the long trip home, she thought. After all, being in space was stressful, and they had been away from home now for over two years.

As she emerged from the hatchway, she saw John through the open door of his sleep pod sitting before his PC, intently studying his monitor.

"Hi, John. You look like you are doing something very important?"

John looked up and laughed.

"Well, for me I am. I'm studying the latest transmission of the Major League baseball standings."

"Oh, and what do they show?"

"Well, my Giants and the Yankees are in first place in their respective leagues. Tokyo is leading in the Asian League, and London and Athens are tied for first in Europe. As usual, things are close in the Latin League, with so many good teams. And also as usual, Cape Town keeps winning in Africa. But you didn't stop by to talk baseball. Come in."

Closing the door behind her, she pushed herself in and strapped herself to his bed.

"John, it seems to me that we are all a bit homesick and irritable. I am wondering if we are all suffering from asthenization. Maybe things would improve if we planned some pleasant contact with each other."

"What do you have in mind?"

"Maybe we could use the July 4th break to relax and unwind and get our minds off of things a little. You could bring your harmonica; we could each sing songs from our homeland, maybe have a contest for the winner; we could do some microgravity dancing; and we could open up one of our "special celebration" dessert packages. What do you think?"

"That would be great! We could also modify the schedule to give ourselves a complete day off from maintenance tasks and watch some video movies together in the dining area."

"Good idea. I will propose this to the rest of the crewmembers."

"Katya, before you go…how do you think Mike's doing? He seems to be very secretive and impatient with everyone. Yesterday, I heard him speaking to Juliette again about CARS, almost blaming her for the incident in May and suggesting that he knows more about the computer system than she does."

"I am not sure what is happening with him. It might simply be a severe adjustment reaction to our mission. I have tried to talk with him, but he refuses, saying everything is fine and that it is all in my head. But I think we are all anxious to get home. The mid-course maneuver next week may help, as a symbol that our return is about half over. Of course, remember the warning made by the training psychologist before launch, that it will be nice to feel that we are nearing the halfway point of our return home at the time of the mid-course maneuver, but that we may also feel despondent to realize that we still have over half of the return time to go!"

"Yes, the glass is still half empty! I know what you mean. But the whole return phase of our expedition has been a bit demoralizing. After all, being on Mars was exciting, with new things to do and new places to explore. Now, we just seem to be biding our time until landing."

"I know. But maybe all of us, especially Mike, will relax a bit after our party. After all, he is an American, and we will be celebrating an American holiday."

"Maybe. But you and I should keep trying to talk with him, and perhaps you could get him to do a little counseling or take some medication. We don't need to have an irritable crewmember disrupting everything."

"OK, John. Is there anyone else you want to talk about?"

"I guess Tolya and Juliette are finding comfort with each other."

"Oh, you noticed?" she said. They both laughed.

"I guess a little fling won't hurt anything, Katya. No one has commented to me about it, and I haven't discussed their relationship with anyone else but you. Do you think it is OK, or should I speak to them?"

"I would leave them alone for now and see what evolves. Tolya seems to have genuine feelings for Juliette, which could be a good thing for him. The relationship might be good for her as well. Her life has been her work, and perhaps there will be some kind of future between the two of them when we get back."

"Katya, are you turning into a *babushka*?" They both laughed.

"Maybe a kind one who has some limits in how actively she arranges for couples to be together."

"How are you doing yourself?"

"I am doing fine, John. I keep busy with things. I sometimes think about Slava and how much he would have enjoyed being on this mission with me. In a way, I think that I am doing it for the two of us."

"Yes, I only met him once when I was training in Russia for a lunar mission, but I liked him and was sorry about his death."

After an awkward silence, she said: "And how about you and your family?"

"I miss them. I always seem to be away during major family events. My daughter had her senior prom and graduated from high school a couple of weeks ago, and here I am."

"Did you communicate with her?"

"You know, during my previous on-orbit and lunar missions, it was much easier. The missions were shorter. Also, I could talk to everyone in my family almost as easily as if they were in the next room. But this damned communication delay really is a nuisance. Receiving delayed A-V transmissions, or writing an e-mail with multiple questions at the end and then hearing back a half hour later…this isn't really communicating. I'll be glad when we are closer to Earth and can speak with people more naturally."

"Yes John, going to Mars takes some adjusting." After another silent pause, she said: "Well, I will leave you to your baseball. I will tell everyone about the party, and I trust you will make the necessary changes in the schedule."

"Yes doctor," he said jokingly, but then seriously commented: "You know Katya, I'm glad you came by. I always enjoy talking with you, and I feel better that I'm not alone with some of my concerns."

"Me too," she said. "As the 'old people' of the expedition, I think that some of the others feel that we are better able to adjust to things and react less emotionally than they do. That puts a real burden on us, and it is good that we can talk with each other about it. Maybe we should talk together more often."

"That would be great, Katya."

As she excited his pod, both of them felt relief. And a sense of support.

17 Coupling

That night, Juliette was restless. The lights had been dimmed and everyone had retired to their pods, but she couldn't sleep.

What am I feeling—loneliness? No, not exactly. Homesickness? No, that's not right either. Longing? Yes, that's the feeling. But for what?

No thoughts came to mind. Then an insight. *How about for whom?*

Tolya popped into her consciousness. Was she longing for him? Yes. Was this simply a sexual desire? No, it was more than that, even though her body began to feel warm. Did she need to see him? The idea of "need" jarred her—she really didn't see herself as a needy person. How could she be needy? But what were her needs? What was missing here?

You must get control of yourself, Juliette, she thought. After a moment, she responded: *Why should I? I will go to him.*

She dressed, opened her door, and seeing no one around floated over to his pod. She quietly tapped on his door, and after a moment he opened it. Only mildly surprised, he smiled, looked out at the empty hallway, gave her a quick kiss, then ushered her in. He had been working at his desk, and he strapped himself down on the chair. He motioned for her to sit on the bed.

"What a pleasant surprise. I was working on mid-course navigation plans but thinking about you, and here you are like dream. Microgravity, maybe it encourages telepathy."

"I couldn't sleep *mon chéri*. I wanted to see you."

Mon chéri? Did I just say that? she thought.

"What a good idea," he said playfully, reaching his hand across to stroke her shoulder.

"No, that is not what I meant. I was thinking, where are we going with all of this?"

"What do you mean?" he stalled, not knowing how to respond.

"I guess what we are doing is fine, the sex is nice, but I feel like we are in a…what is the English phrase…a 'goldfish bowl.' I am sure people suspect our relationship, but I am not sure what our relationship really is. I have been alone and focused all my life, and I do not know what I feel now, why I am involved in this way. Maybe it is the mission, the isolation, but maybe not. How do you feel?"

This was a serious question. He took stock of the situation: his feelings were growing for Juliette; she was on his mind a great deal; he looked forward to seeing her and felt frustrated when he couldn't kiss her in public; he longed for her the nights they retired into their own pods after a busy day. Was this a fling brought about by their situation? Or was it more?

"I do not know," he said honestly. "I care for you and want to be with you, but we are not in normal situation here. We have isolation, confinement, limited privacy, boredom, danger…I do not know how to feel about anything."

"Me too, and there is no place to run."

Struck by the similarity in their feelings, he responded: "Yes. On Earth, you can have fling, then go away to settle things down. Here, that is not possible."

"But is that what you want to do? Just have a fling?"

"No, I confess not". He added: "When I was boy, because we moved so much from air base to air base, I learned not to get close to people, since we could be gone in few months. That was good thing, since my life was not complicated. I became expert in quick relationships." He thought further for the right English words. "They were passionate, intense, but also safely temporary. But that is not case here…"

"I know, I feel the same way. I think I used my work to run away from people."

They were silent for a moment. Then Tolya surprised her by saying: "I want to try this for a while, maybe even after we go home, and see what happens. Neither or us will fly in space again, but both of us will be world famous. Maybe we can cut deal where we advertise apartment complex in Paris or Moscow and get free apartment together in return."

She laughed at this possible perk from the fame they would experience after the expedition ended. But then Juliette thought to herself: *Mon Dieu! What is he proposing? Can I do this? Can I commit to something like this after the mission? Can I commit to anything even now.* She became frightened.

"I don't know. I care about you, but…"

"I care about you, too. We must try to form life together after we get home, to see how we get along."

"We have been away so long, it is hard for me to think about any kind of life after the mission," she half-joked to break the tension.

"Yes it is hard," Tolya responded. "Well, we have several more months to go before landing. We have time to think plans over."

She nodded in agreement, undid the bed strap, and floated over to him. They held each other in silence, then nuzzled, then began kissing. Both slept well later that night.

18 Asteroid

At breakfast the next morning, the crewmembers were discussing the July 4th celebration when a buzzer rang on the control counsel. Tolya floated over, checked the monitor, then quickly positioned himself in the Pilot's chair.

"Commander, emergency back-up radar system is tracking something coming our way. It is moving fast and has been programmed to intercept us in eight minutes."

Surprised, John quickly floated over and strapped himself down in his chair before the control console.

"What do you think it is?"

"It is solid body, about three hundred forty meters across."

"A rogue asteroid?"

"Most likely. There is no coma, and the reflectivity does not suggest cometary nucleus made of dirty ice. It is very dark. I believe asteroid is most likely explanation."

"We don't have much time. Why didn't CARS pick it up on the primary system?" He glanced over at Juliette, who had gone over to the central computer terminal.

"I am not sure," she responded. "The emergency system has its own independent power supply and is separate from CARS, but I do not know what happened to the primary."

"Tolya, is it still on a collision course?"

"Yes, Commander. We must initiate evasion protocol."

"Everyone to their emergency stations."

There was a flurry of activity. Mike, Jango, Katya, and Juliette headed for their acceleration chairs. After strapping themselves in, they all swung the metal arms holding the individual computers down in front of them and proceeded to carry out their assigned duties. Mike and Jango examined the status of the various operational and engineering systems. Katya tracked the life support systems. And Juliette began to monitor the CARS programs and subsystems.

"How is CARS behaving, Juliette?"

After a moment, she reported: "Everything seems nominal. All systems appear to be working properly."

"I see same thing," reported Tolya after glancing at the control console.

"OK. Tolya, program an evasion course."

The Russian pilot initiated the evasion sequence on the central navigation module, and the course screen lit up. It showed a graduated green line pointing upward that was being intercepted by a red dot coming in from the left side. Leading away from the green line was a curving blue line that deviated to right. This was the planned evasion course.

"Programming completed, Commander."

"Copy that. Initiate thruster sequence."

Tolya moved a toggle switch on the control console forward, but nothing happened.

"Commander, thrusters fail to respond."

"What? Juliette, what's happening."

Glancing at her keyboard, she responded: "I do not know. CARS shows everything nominal."

"Try again, Tolya."

The pilot flipped the toggle forward and backward several times, to no avail.

"Nothing again, Commander. And thruster activity lights are not on. We are not able to change course!"

19 Interlude

It became aware that this was a serious matter. The asteroid was threatening the lives of all of them, including the progeny. It realized that It must not continue to interfere with the propulsion system, or all of them would die. It quickly took steps to release all of the blocks and allow the thrusters to activate. It decided that it was time to stop testing and to activate the Plan.

Perhaps the best time would be during the holiday celebration. They would be relaxed then, their guard would be down. Yes, that would be the best time, It thought. Patience until then!

20 Course Correction

"Wait, Commander, thruster lights just went on. I will activate course correction again."

Tolya pushed the toggle forward. A vibration shook the ship slightly as the thrusters fired according to the activation sequence. John looked out the forward window and saw the stars begin to shift direction. Looking down at the control panel, the course screen began to reflect the ship's movement, as the green line began to deviate toward the blue line. Slowly at first, the movement began to gain speed until the two lines merged.

Looking out the window, John spotted a rapidly moving object to the left. It periodically flashed with the reflected light of the distant Sun as it tumbled in space. It was slightly oblong in shape, and John thought he saw a small crater near the center. It approached the ship, moved parallel for a while as the ship continued to veer right, then whisked forward and to the left into deep space.

"Well, we missed it. Or should I say it missed us. Good thing you reacted to the thruster lights, Tolya."

John spun his chair around.

"What happened Juliette?"

"I don't know. I show all systems nominal. CARS recorded the proper co-ordinated thruster firing and the planned course change that we experienced. Let me check out a few things."

As she examined her keyboard, John thought about what had just happened. *Another glitch with CARS. What's going on? This time we escaped with our lives, but next time we may not be so lucky.*

"John, looking at the time sequence, I show that there was some sort of thruster control block in place when Tolya first tried to change course. This would account for the lack of movement and the dead thruster lights. Then, for some reason, the block was released and the thruster system became operational again. I do not know why."

"Was it something Tolya did?"

"No, he simply activated a system that became operational."

"Do you think this was linked to the thruster problem Mike discovered in May?"

Over to the side, John noticed Mike smirking as he fidgeted in his chair.

Juliette pushed some buttons on her keyboard.

"No, the thruster pressure warning system and the course correction system are not related. In fact, I show that the thruster pressure warning system was completely operational throughout the entire course correction sequence."

"So what happened?"

Frustrated, she responded: "I do not know! It was like CARS was trying to block our course control ability, then changed his mind."

CARS is not a person! John thought. *But whatever is messing with us may very well be.*

"Keep an eye on CARS and let me know if something new turns up. For now, though, it seem that the crisis is over. All of you—go back and finish your breakfast."

Everyone but John unstrapped themselves and headed for the dining area. John continued to examine the evasion course to see how it had affected their trajectory back to Earth when he became aware of a presence next to him. It was Mike.

"Commander, Juliette just can't handle CARS. The system is not functioning correctly. It's going to kill us all. We are in for a fall. Let me fix it. I know how everything works in this ship. I can take care of it."

"But Juliette is an expert…"

"She's over her head. It's a basic engineering problem. That's what I do."

He paused, chuckled, then continued.

"I can take care of anything."

John puzzled about how to respond to this comment. He decided to take a direct approach.

"Mike, I don't know what's going on with you. You seem to have it in for Juliette. But you're stepping outside the bounds. I know you're the chief engineer for the mission, but you're exceeding your expertise. Juliette is in charge of CARS operations."

Reddening in the face, Mike sputtered his response.

"She's got you seduced as well. First Tolya, now you. You're not giving me enough credit. I can fix anything on this ship. You're endangering the mission by trusting her. You need to trust me. I was top in my class. I know systems. I know computers. I can save us all!"

"Mike, I think you need to talk with Katya about your feelings…"

"I don't need to talk with her. What does she know? She just listens to our hearts and counts our pulses. I know the pulse of the ship. I don't need her. I don't need any of you!"

Mike used the edge of the control panel to push himself away and floated to the hatchway leading to the upper deck. Jango and Tolya looked at him, puzzled, as he floated by. An incredulous John wondered what to do next.

Katya came over to him.

"Well, this certainly has been an exciting morning."

"Yes it has. But I can tell you one thing. With all this stimulation, I think we have all been cured of asthenization!"

21 July 4th

After the events of the previous day, July 4th started out very well. Everyone but Mike came down on time for breakfast, and they all were upbeat and ready to have a relaxing day. After eating, they agreed that it was time for their singing contest, but since "Yankee Doodle Dandy" was first on the agenda, John wanted Mike to join him in the chorus. He also wanted to assess the engineer's emotional state. He called up the hatchway, and Mike yelled that he would be right down. In the meantime, John played his harmonica and the crew tried some microgravity dancing, which largely consisted of spinning, somersaulting, and flailing their arms to the beat of the music. It felt silly and looked silly, but everyone was in good cheer. There was no trace of ennui on this occasion.

The decision to go to Mars under conditions of microgravity had been hotly debated early in the planning of the expedition. Life scientists agreed that negative physiological effects on bone, muscle, balance, and the cardio-vascular system could occur by prolonged microgravity. The question was: what to do about it? One solution was to have a small centrifuge on board that would allow the crewmembers to take turns spinning under "1 g" Earth grav-ity, or even hyper-g forces. However, mass and space considerations involving a possible centrifuge created engineering and energy problems. In addition, there was no evidence that a few hours under partial gravity would be enough to offset the effects of microgravity during the long flights to and from Mars. So, the planned centrifuge was abandoned.

Another suggested solution was to eliminate microgravity completely by creating permanent artificial gravity. But how? Advocates argued that a teth-ering system could be devised to connect two vehicles going to Mars, one manned and the other carrying supplies. These could be set spinning around each other at a long distance and at a speed that produced a centrifugal force of 1 g in the manned vehicle. In a sense, the crew would be taking the Earth's gravity along with them. Those opposed to this plan pointed out that spin-ning people around a common center of gravity could cause inner ear distur-bances. In addition, the technology needed to create a stable tethering system was untested, and doing so would increase the mission costs prohibitively. So this plan was abandoned as well.

The final solution was to instill an elaborate exercise regime for the crew-members whereby proper physiological tone could be maintained to mini-mize the effects of microgravity during the outbound and return phases of the mission. It was hoped that the 38 % Earth gravity of Mars would help as well during the crew's stay on the Red Planet, although supplemental exercise would still need to be performed. So far, the exercise regimen seemed to be working in keeping the crewmembers' physiological state intact.

But there was one positive effect of microgravity: it could be fun. Here was the crew, floating and spinning and having a good time at a party. None of them had yet tired of this aspect of being weightless in space.

"Sorry to be late, people, but I've been up all night working on a way that we can get home faster," Mike said enthusiastically as he emerged from the hatchway opening carrying a large rolled up sheet of paper. "Actually, I've been working on this for several days during my leisure and sleep time. You know, we really don't have enough time in the day to waste it on sleeping. I think the mission planners should have recommended fewer hours for sleep. Maybe two or three hours a night—that should be enough. But then, we don't have to depend on those jokers. We can control our own schedule. Can we do this, John? Anyway, where was I? Oh yes…"

Mike waved the roll of paper and in the process started to spin in the microgravity until he grabbed on to the side of the hatchway door to steady himself.

"According to my figures, by burning our main engine at full capacity for 2 minutes and 13 seconds, and then activating the yaw, pitch, and roll thrusters according to the sequence in my calculations, we can produce sufficient acceleration on a new trajectory that will get us home 9.3 days earlier. Although our angle of entry would be close to the tolerance level, we could depend on our heat shield to land us in some wheat field."

He laughed at the rhyming of the words "shield" and "field."

John and Katya looked at each other, stunned, then looked back at Mike. He was wearing a red M.I.T. tee shirt, his underwear, and gray socks. His hair was disheveled, and he was unshaven. He began gesticulating with his arms, which caused him to begin to flip over, until he again grabbed the hatchway door.

"This is a great plan," he continued. "It will be an operational breakthrough for future trips to Mars, and even beyond." He started to back through the hatchway.

"Mike, wait," said John. "Your plan sounds interesting, but wouldn't it be dangerous due to the steep angle of entry into the Earth's atmosphere?"

Mike suddenly became angry. "What are you, some kind of 'girly-man'? How did you get to be Commander, anyway? My plan will work. It's revolutionary."

"But Mike…"

"Don't 'but Mike' me, John. You're just jealous and afraid that I'll take over from you. What do the rest of you say about this? Shouldn't I be in charge? Let's have a revolution," he said laughing.

Realizing that the American engineer's condition had seriously deteriorated, Tolya and John eyed each other, then slowly floated over toward him. Jango and Juliette moved back to give them room, and Katya edged toward the medication cabinet.

Tolya said: "Mike, John is still in charge, and we want to discuss your ideas, but not here. Why don't we go to exercise table, where you can spread out plan and show us."

He motioned in the direction of the table, which was rimmed by the restraining straps used to hold the crewmembers in place when they exercised.

At first, Mike complied and began floating over, but nearing the table, he stopped himself by grasping its edge.

"You know this is my plan. I don't know if I can share it with you. You all want my things. Keep away from my things."

John spoke: "But Mike, how can we consider your plan if you don't let us see it? We won't take it away from you. Just spread out the paper on the table."

Eyeing the restraints, Mike turned toward John.

"Wait a minute, what are you gonna do? Are you trying to stop me? I will stop you first! CARS and I will stop you. I don't need any of you. CARS will help me. CARS will understand the plan. He's my equal. He and I will get back safely and will tell everyone about this plan. They will use it to go to the outer planets."

Tolya was the first to reach Mike and grabbed his left arm. As the two of them started to spin, Tolya grabbed the table for support, and John grabbed Mike's other arm.

"Let go of me, you Commie! You are in cahoots with John to keep me from being Commander. Today is our Independence Day, not yours. We have the Declaration of Independence. I declare myself Commander!"

Katya floated over and injected Mike's shoulder with a syringe full of tranquilizer. He continued to rage for a few moments, then began to calm down enough for Tolya and John to pin him to the table and for Katya and Juliette to strap him down. Jango remained silent in the corner, looking fearful and surprised.

"What's going on here, Katya?" asked John.

"It looks like a manic episode to me," she responded, looking down at her new patient as she took his pulse.

"But how can this be? Bipolar disease is genetic. It runs in families. Why wasn't the selection committee alerted? Mike is 38 years old. If he's prone to this disease, why didn't he have an episode before?"

Katya looked up at John.

"Although rare, a first manic episode can occur in the late 30s. Remember, Mike is adopted, and no one knows about the psychiatric history of his genetic parents. He was given a waiver for this omission since he was such a top-flight engineer. During previous shorter missions, the psychological reports on Mike noted him to be emotionally excitable at times, but this was thought to be part of his personality. He performed his duties extremely well, and he generally got along with his crewmates. In our longer mission, we have been under a lot of psychological pressure, and I guess this was just his time to have a first episode. He apparently has been working non-stop in his pod, even into the night after the rest of us have gone to sleep."

She paused briefly as she adjusted Mike's restraints, then continued.

"I did not put this all together. His cranky intrusive behavior and wild clothes should have alerted me as to what was going on. I should have suspected that he would be having a psychotic breakdown."

"Damn, what a time for something like this to happen," said John. "What should we do now?"

"We should keep him strapped down until we can get him properly medicated, then we'll see. Hopefully, he will calm down soon enough to be able to eat and drink and move around. We don't need him getting physically sick or having some sort of cardiovascular problem from stress or lack of exercise."

She noted the abrasions on his arms where he had earlier pulled against the straps.

"Nor do we need him to hurt himself. This is not a proper seclusion room. Maybe Mission Control will have some idea of what we should do."

As the expedition physician, Katya had fretted about what would happen should someone become psychotic or suicidal during the mission. Since the crew had been carefully screened psychologically, it was believed that the odds of a true psychiatric emergency were small. Although there were protocols for treating adjustment reactions and minor depression, routine medical issues, and accidents requiring minor surgery, there was very little written down about ways to seclude and restrain a seriously mentally ill crewmember. There was no dedicated seclusion area, and the major restraining apparatus on board was the exercise table.

Katya believed that this lack of planning reflected a lingering prejudice that permeated space agency personnel and the astronaut corps, some members of whom viewed psychologists and psychiatrists with a skeptical eye. After all, couldn't they ground a crewmember during training if they thought that something was amiss? Much effort had been spent on selecting the *MarsExplore* crew, and no one wanted to see a crewmember grounded at the last minute. Also, it seemed to Katya that some mission planners were engaged in denial ("there is no way that this will occur, not with this great crew") and magical thinking ("if we don't dwell on it, it will never happen"). Asthenization or stress reactions were OK—they could happen to anyone. But a psychotic reaction like acute mania, with the eyes of the whole world looking on, could be a public relations nightmare.

John seemed to be thinking the same thing. "Katya, maybe you should wait a while before contacting Mission Control," he said. "Mike is quieting down. Maybe he's reacting to something toxic that will go away or can be treated. By waiting a bit, we might have a better report to give to Mission Control. I hate to get them all riled up before we have some answers."

"OK, John, that makes sense," she said.

"Mike mentioned something about working with CARS to stop us," commented Tolya. "Do you think that he is behind thruster pressure issue? Maybe he altered program."

"But he also discovered and reported the problem," volunteered Juliette. "Why would he sabotage something and then report it to us?"

"I don't know," replied John. "Who knows how long he's been getting sick. Maybe he thought that by telling us we would think he was some sort of hero. He wanted to take over my job and show you up. He seemed to want to prove that he was better than all of us. Maybe that was his motive. People with mania get grandiose ideas, right Katya?"

"Yes, that certainly could happen. But now we have him restrained, so if he was the culprit, he won't be doing any more damage for the time being. I will run some blood and urine tests to rule out a drug or toxin as the cause of his acute episode. Also, I will examine him to see if there is some other medical issue going on. Jango, since you were cross-trained as a medic, you and I will need to set up some sort of schedule for monitoring Mike and making sure that he is well medicated. It sometimes takes a while for a person with an acute manic reaction to calm down enough to move around unrestrained and to interact with other people. In the meantime, we don't want him pulling on his straps and breaking a bone or otherwise injuring himself. Maybe the rest of you can take a turn with Jango and me in watching Mike." They all nodded in agreement. "I will also modify our exercise program—Mike has priority right now for the table."

****

Mike continued to be agitated for the rest of the day. By the next morning he had calmed down enough for Katya to examine him and to run her tests. Nothing was found to support a toxic or medical cause for his episode. She and John finally told Mission Control of the event, and they responded by calling in some psychiatric consultants who were familiar with the treatment of bipolar disorder. They gave Katya their advice on the best way to manage the situation, which was in keeping with her clinical ideas.

For the first two days after the Wednesday holiday, Mike's condition alternated between medication-induced sleep and loud insulting outbursts, which distracted the crewmembers from their work. It was especially difficult during the day, since the exercise table was on the same deck as the primary work areas in the command center and the laboratory, so Mike's distress was apparent to everyone. Mercifully, his outbursts stopped whenever he received more medication. He also received fluids, vitamins, and sugar intravenously through a special positive pressure delivery system adapted for use in micro-gravity. Katya was relieved to see it work so well.

Katya developed a schedule whereby someone was watching Mike at all times for his own safety. The crew shared this responsibility with good humor and a sense of camaraderie. By Saturday, during periods when the medication had calmed him down, he could be partially unstrapped and feed himself with assistance from Katya and Jango. But at night, it was felt to be prudent to keep him restrained to the table.

22 Awakening

Early on Sunday morning, the 8th of July, John was awakened by a headache. *Damn, I really wanted to sleep,* he thought. *It's been a tough few days, and I want to be alert for the mid-course maneuver later in the week.* Feeling a bit flushed, he turned over in his bed. He became aware of his heart pounding and his muscles twitching.

Am I having an anxiety attack? Boy, we sure don't need another psycho on board!

He then bolted upright, feeling a bit dizzy.

Wait a minute…I felt like this once in an altitude chamber training session when they raised the CO_2 *level!*

He unstrapped himself and floated over to his PC. He called up the life support numbers, and he saw that the CO_2 level in the air was over 3 %. Pulling on his shorts, he exited from his pod and made a quick circuit through the upper deck, banging on each door and yelling: "Air emergency! High CO_2 level! Head for the lab!"

He then floated down through the hatchway to a cabinet in the laboratory that housed two emergency air canisters. Breaking the cover door seal, he pulled one out and took some deep breaths. He glanced over to the exercise table, seeing only Mike strapped down, apparently asleep but breathing hard.

Who is on duty to watch Mike tonight, he thought. *Jango. Where is he?*

Just then Katya floated down, looking apprehensive.

"Katya, take a couple of puffs of this, then grab the other air canister. Give Mike a couple of breaths, then go back up to rouse everyone else. Our CO_2 level is over 3 %. I'm going to the lower deck to put on my spacesuit. Send people down after you give them a couple puffs of air. We all need to be suited up until we can get our atmosphere back in order."

She nodded and headed for the second canister as John floated down to the lower deck. As he emerged from the hatchway opening, he saw Jango, fully suited, in the corner. John went over to his designated suit, took two more breaths from the air canister, began dressing, and yelled: "Jango, what happened?"

"I began feeling dizzy and short of breath, so I immediately came down here to put on my suit. I was just going to go up and warn everyone."

"Go up now and help Katya. Send people down here to put on their suits as soon as they can."

Jango went up, and a minute later Juliette appeared at the opening. She was breathing hard and looking worried.

"Here, take a couple of breaths from the canister, then put on your suit."

As he sealed his helmet, John was grateful for the new spacesuit that was specially designed for this mission. It allowed one person to quickly put it on. He activated his suit's internal air supply. It felt good to take several breaths of air—even the suit air smelled good. Leaving his canister with Juliette, he floated back up the hatchway, seeing Tolya coming down from above.

Good, everyone is accounted for. "Tolya, go down and get suited up. Katya, give your canister to Jango, who can attend to Mike while you go down as well."

As everyone followed orders, John went over to the control console. He called up the life support panel on CARS and saw that the CO_2 level was 3.57 %.

How can this be? The carbon dioxide scrubber is not working. Why didn't CARS adjust it? Why was there no alarm?

A suited up Juliette joined him at the console.

"Juliette, what's going on?"

She called up a program, made some entries, and called up another program.

"There is an instructional program deactivating the scrubber. In fact, it has been put in reverse and told to release stored carbon dioxide into the air. Furthermore…"

She made some more entries.

"The warning alarm has been deactivated, much like what happened before with the yaw thruster and asteroid events. I don't understand."

"Can you fix it?"

"Yes. I need to reprogram the instructions. First, I will tell CARS to have the scrubber accelerate carbon dioxide absorption from the air, then put the absorption rate back to normal. I will also reprogram the CO_2 detector and reconnect it to the alarm system. The air should be breathable in about 20 minutes."

"Good, we will have enough air in our suits and in the emergency canisters to wait things out."

****

Ten minutes later, five suited figures floated in the mid deck. Katya and Jango were giving Mike puffs from the air canisters, and John, Juliette, and Tolya were huddled around the control console.

After ten more minutes, Juliette gave the all clear signal. Everyone waited while John checked the computer readings himself, then carefully took off his helmet, sniffed a couple of times, and pronounced that things were back to normal. They all went down to the lower deck to take off their suits, first John and Juliette, then Jango and Katya, and finally Tolya. After this was done, they all gathered in the command center on the mid deck.

"That was a close call," said John. "Let's review what just happened. First, CARS again malfunctioned, or was made to malfunction, so that the CO_2 level rose without our being aware of it. Second, the warning alarm did not go off. Third, the problem occurred at night, when we were asleep."

"Not all of us, Commander," said Tolya. "Jango, you were supposed to be awake watching Mike. Did you see anything suspicious? Did you do anything to cause this?"

Jango became flushed. "I already told John that I was aware of some breathing problems and immediately went down to put on my spacesuit."

"But why didn't you warn the rest of us?"

"I thought that I would first secure my own air supply, then go upstairs and attend to anyone needing help."

"Did you think about the emergency air canisters?" asked John.

"Yes, I would have procured one of them on my way up to awaken the rest of you."

"Assuming we could have been aroused. Anyway, Juliette, what happened to CARS?"

"Something like what happened before. But whatever it was, it was a recent change. The carbon dioxide system has been working nominally until now. I just checked it last Friday, and even the alarm was working."

"Did someone alter the program since then?"

"Probably. But we know that Mike was not involved, given that he was pretty much incapacitated."

Everyone looked around at everyone else.

"So maybe we have a saboteur, maybe not. But could CARS have done this?"

"What do you mean, John?"

"I mean, could the computer have altered itself in some way? Could it have changed some of its own programs?"

"John is asking if it tried to kill us," Tolya said, to no one in particular. Then looking at Juliette, he said: "Could it be thinking for itself?"

She thought for a moment, then responded: "It is a very sophisticated machine, with more memory and operating speed than any other computer built before. People have speculated that such a machine could be conscious, could have motives, even feelings."

John commented: "Like in the movie '2001', where HAL, the computer, malfunctioned, then read the lips of the two crewmembers who were talking about turning it off. It then tried to kill them before they could carry out this action. Is that what you're saying, Juliette? If so, what would be the motive for CARS to do such a thing? Could it think that we are trying to turn it off for some reason?"

"We really haven't considered it," she responded.

"I have idea," said Tolya. "We are in last phase of mission, and CARS will have no purpose after we land. It may be deactivated, turned off by World Space Council. It would die. But by killing us, it could fire thrusters and alter ship's course and continue flying through space. It would be in operation for eternity. Remember, one thruster already was altered, so such a thing is possible."

"That sounds like another science fiction movie," interrupted Katya. "There must be a more reasonable explanation than sabotage by a thinking computer worried about being turned off."

"Whatever the reason, we need to be more vigilant," said John. "Juliette, see if you can determine exactly when the programs were changed. Also, from which computer terminal. We all have PCs in our pods that link up with CARS. If one of those machines was used, perhaps we can find out which one. In the meantime, maybe we can program in a warning system that would trigger an alarm if anyone or anything tries to change something in the computer. We need to review our procedures for dealing with emergencies like what happened tonight. Jango should have called an alarm and gone for the emergency air canisters before he went to put on his suit."

Everyone but Jango nodded in agreement.

"I think we need to have more frequent checks of our life support systems. Also, our propulsion system. We can't be caught off guard again."

More head nodding.

"All right. Dismissed."

As the crewmembers began to move toward the dining area, John held Katya back.

"What do you think?" he whispered.

"I do not know, John. There must be a more reasonable explanation for what has happened. Something we have not thought about. After all, CARS has performed beautifully for over two years in all our vehicles and habitats, and we have all worked with each other for nearly four years of training and this mission. Even asthenization would not push someone into murder."

"Well, let's continue to be vigilant."

"Agreed. John, one more thing. If it is the computer, what will happen when we try the mid-course maneuver?"

"What do you mean?"

"Well, if CARS does not want us to go home, what will it think about a burn aimed at doing just that? Will it refuse our instructions, or will it send us on a different course?"

John thought a bit about what she just said. "I guess we'll just have to see what happens."

She nodded, and they joined the others at breakfast.

23 E-mails

To: john.c.wood@MarsExplore.gov (PERSONAL AND CONFIDENTIAL)
From: charles.j.szegel@nasa.gov
Subject: Crewmember Issue
Date: Sunday, 8 July 2035, 1905 hours CDT

John,

I have some bad news to tell you that involves one of your crewmembers, Ekaterina Malenkova. I just heard from Boris in Moscow that yesterday Dr. Malenkova's sister, Valentina, was involved in a hit and run accident while crossing Nevsky Prospect in St. Petersburg. The driver was in a red sports car and drove away after hitting her. There were witnesses, and the police are looking for him. Unfortunately, Valentina was dead on arrival at the hospital.

We just finished an emergency meeting here where we reviewed the policy of the World Space Council about dealing with such an incident. The policy states that you are to be informed first in your role as the Mission Commander. Use your judgment as to when to notify Dr. Malenkova. It should be as soon as possible, but not during a mission critical activity.

Please express our condolences to Dr. Malenkova. Tell her that both Dr. Vadim Smirnov and I are available here should she wish to speak with one of us in private about the event. We are happy to provide support in any way necessary.

Charlie

Charles J. Szegel, M.D.
Flight Surgeon
MarsExplore Expedition Program
NASA Johnson Space Center
2101 NASA Parkway
Houston, Texas 77058

****

To: charles.j.szegel@nasa.gov (PERSONAL AND CONFIDENTIAL)
From: john.c.wood@MarsExplore.gov
Subject: Crewmember Issue
Date: Sunday, 8 July 2035, 2013 hours CDT

Charlie,

I received your e-mail about Katya and the death of her sister. I will tell her the news tonight. Even though we have a lot going on here right now (what with Mike's illness and our impending mid-course burn), I don't see any reason to wait. The sooner she knows, the better.

Katya is a strong and very professional person. I am sure that she will be able to deal with the news and still carry on with her responsibilities. I will get back to you if she wants to speak with someone there. Of course, we all are available here to support her. Even though she is our crew physician and mental health expert, I think we crewmembers are sensitive enough to help her through the crisis. If we need any expert advice from you all, I will be sure to let you know.

Take care. Give my best to Boris, Vadim, and everyone in Mission Control.

John

****

To: charles.j.szegel@nasa.gov (PERSONAL AND CONFIDENTIAL)
From: john.c.wood@MarsExplore.gov
Subject: Crewmember Issue
Date: Sunday, 8 July 2035, 2019 hours CDT

P.S.,

Valentina was Katya's only living relative. I hope they find the bastard who killed her soon and string him up! What a coward he was to drive away. He was probably drunk.

Keep me informed as to what happens with the police search. I will pass the news on to Katya—she will want to know.

John

24 Bad News

I guess now is as good a time as any to tell Katya, John thought as he turned off his PC. He felt badly for her, but with the mid-course maneuver just two days away, all of them would need to be in strong emotional shape by then.

But damn! The timing is still lousy, what with Mike's breakdown and the problem with the carbon dioxide. When it rains, it pours.

He floated over to Katya's pod. It was still early, so he was sure that she would be awake. He knocked on the door. She opened it and gave him a big smile.

"John, what a surprise!"

"Katya, I have some bad news that I need to talk to you about."

She suddenly got serious. "By all means, come in."

He entered her pod, she closed the door, and they strapped themselves down on the chair and bed, respectively.

"Katya, earlier tonight I received an e-mail from Charlie Szegel in Mission Control. He heard from Boris in Moscow that your sister Valentina was involved in a traffic accident yesterday and has died."

"Valya! O my God! Dead? What happened?" Tears began to well up in her eyes.

"Hit and run driver. The police are looking for him. Katya, I'm terribly sorry."

"Oh my God." She made the Orthodox Sign of the Cross and muttered something in Russian. "And she had been so happy lately. She wrote me that she had just found a new apartment. Since her divorce, she has been despondent, and this was a very positive thing for her. I haven't seen her in years. We only communicate using e-mail." Tears streamed down her cheeks.

John reached for her hand and held it as she sobbed quietly. After a minute's silence, he continued.

"Charlie said that he and Vadim are available if you want to talk with them about anything. I am also following up on any news about the hit and run driver."

"Thank you. I will be all right. I suppose it really does not matter who the driver is. It will not bring Valya back. You know, since I moved to Moscow after my marriage, I had not communicated with her very often—too busy, I guess. It got worse when we trained for this mission. And of course, I have been away for over two years. I only wish that I had seen her more often."

"I think we all feel that way about our families and friends."

"Maybe this is the biggest problem with space travel—losing contact with people you love on Earth."

"Yes, it is an occupational hazard. But in a sense we are your family. Is there anything I or anyone else here can do?"

"No, I will be fine. I just need a little time to think about things. Poor Valya. She was really getting her life back in order."

"Would you like to be relieved of duties for a day or two? I'm sure that Jango and the rest of us can cover things pretty well."

"No thank you. It will help me to be involved. Things can be isolating enough up here without me isolating myself even more by staying alone in this small pod."

"When should we tell everyone else?"

"Maybe tomorrow, at breakfast?"

"Do you want to tell them, or should I?"

"Maybe you should—it will be a sort of formal announcement."

"OK. Is there anything else you need?"

"No thank you, John."

He unstrapped himself and gave her a kiss on the cheek, intending it to be brief before he left. To his surprise, she grabbed on to him and hugged him, sobbing quietly. It was a bit awkward, since he started to float away and had to secure himself in position by holding the bed frame with his hand. But neither of them seemed to notice.

25 Revelations

The next morning, after Mike had been tended to and everyone had gathered for breakfast, John announced the death of Katya's sister. They all expressed their condolences. Juliette floated over and hugged Katya. People volunteered to help her in any way that she would like. She thanked them all for their concern but said that she would be all right.

After a moment's silence, John spoke again, in part to see how everyone was feeling about the event.

"You know, it's difficult not to think and worry about family and friends back home. People talk about everyone worrying about us astronauts, but it works both ways."

Tolya responded: "Yes, my mother is in nursing home for dementia. Her health is not good. I expect any day to hear that she died."

"My father is also sick, with heart trouble," said Juliette. "When I last e-mailed him, he said he was 'holding on' until I returned from this flight. He and my mother are very proud of me, but I worry about them. I will be happy when this mission ends and we get back home. I'm tired of being in space."

John was a little surprised by this revelation, since Juliette had always impressed him as being totally dedicated to her work. *But then*, he thought, *we are all a bit antsy to get home.*

"What about you, Jango? You have family back in China."

"Yes, I try to e-mail them when I can. Of course, it is difficult to contact my parents on the farm unless they go to my brother's house and use his computer. I do contact my wife and son in Shanghai. Everyone understands the importance of my mission to my country and to society. I know that I have made a contribution, and I will continue to do so after I return, perhaps more than people think."

John was puzzled by this statement and by Jango's failure to make eye contact with him, but he shrugged this off as just another way the geologist was using to isolate his feelings.

I will never understand him completely, John thought.

Turning to the group, he said: "My wife's mother has been sick. She is very concerned about this, plus dealing with our last child leaving home for college, plus taking care of the home front. And here I am on an exciting enterprise that takes me away from my family for years. I sometimes feel like a deserter."

"Yes," said Juliette, "it is easy to feel guilty about things and wonder about the choices one has made. I never thought that I would say this, but lately I have begun to evaluate my life and realize that being in space is not the only thing there is in life. I was looking at the Earth through the telescope last week, and it was this beautiful, fragile, living jewel in the dark sky. The people I have cared about growing up are all there."

"I wish I would have taken a little more time to visit my sister," said Katya. "We were close as children, then as we grew up, we each went to school and got married and went our own ways. Living in different cities made it more difficult to see each other. But it was not impossible. We just didn't take the time. Now, I will never see her again."

There was silence in the room. Then Tolya spoke.

"For me, it has been worth it to fly to space. But after expedition, I don't know what I will do. Maybe I will become cosmonaut bureaucrat. How exciting."

They all laughed.

"I've had the same thought," John admitted. "I want to stay associated with the space program, but I love to fly and will really miss that aspect of my work. What a dilemma: when we are here, we long to be home; when we are home, we can't wait to get back into space."

Again, thoughtful silence, until their reverie was broken by a loud moan from Mike.

"I had better go over and give him some more medication," said Katya. "But before I go, I just want to say that I appreciate the concern all of you have expressed for me. Thank you. I see you as my family right now."

She then floated over toward Mike.

"OK, everyone, I guess we had better get back to work," John said. "We have our mid-course maneuver tomorrow to think about, and the day's activities to perform."

They all went their separate ways. However, the crewmembers of the *Mars-Explore* Expedition were unusually preoccupied with their thoughts that day, and they sent out a record number of e-mails to family and friends back home.

26 Recovery

By Tuesday morning, Mike was much better. Although he was still a little groggy and his speech was slurred from the side effects of the medication, his tangential thinking and intrusiveness had decreased. He was able to feed himself and to interact better with his crewmates. He no longer needed to be restrained, except for two straps that kept him secured to the table so that he would not float away when he slept.

"Mike, you look great today," Juliette said as she brought him his breakfast.

"I'm getting there. I sometimes still think that I should be commanding this expedition and that I have a special relationship with CARS. Katya says that these are just delusions of grandeur and part of my breakdown. I guess maybe she's right—she's the doctor. What a lousy few days."

"It was quite an experience for us as well. But the main thing is that you are better."

"I never expected something like this to happen. I always thought that I was smarter than most people, but nothing like recently. I really believed that I had unique—even supernatural—powers of thought. On the one hand, I felt confident and special, but on the other hand, I felt angry and impatient with how slow everyone else was and how they just couldn't keep up with me. Only CARS was capable of doing this, I thought, and I felt a strange kinship with it. Now, I just feel ashamed. And Juliette, I'm sorry I was so mean and insulting to you."

"That is all right. I know it has been a difficult time for you as well."

"But with your engineering background, maybe you can help us deal with CARS problem," Tolya commented as he floated into the mid deck from the hatchway. "Maybe you hear about problem with carbon dioxide? CARS didn't warn us again, like with thruster."

"Yes, Katya told me. There is talk that CARS caused all this. I can't believe that a computer system can be conscious or evil. After all, it's just a machine, although one that's very sophisticated. Maybe there's some mechanical flaw, a hardware problem. I can look into it."

He started to get up, but the straps held him to the table.

"You certainly are making progress, Mike," Katya commented as she float-ed over from the laboratory. "But you can't rush things. You need to keep taking your medication. And Mission Control wants us to keep you off duty for a while. Maybe you will be in shape to resume your duties by next week."

"I think inactivity will make me worse. I need to work, Katya. I'm not good at hanging around. I feel a little embarrassed and depressed by what happened, and I need to be busy to get my mind off of these feelings. My life needs to return to normal."

"One step at a time, Mike. Maybe you can continue with your project of taking pictures of the Earth through the telescope."

"Funny, but I've been thinking a lot about the Earth and going home while I've been lying here. I'm feeling more melancholy but also eager to be back. Maybe this is part of my illness, but I feel I have a new understanding of our expedition and its relevance for the Earth. Here we are, finishing an historic mission to another planet, my dream since joining the astronaut corps, and yet I'm beginning to appreciate how wonderful our home planet is and how much I want to be there, to help make it more special. In a way, having been on Mars makes me more of an Earthling."

Katya reflected on what he said. Yes, his psychotic break certainly had shaken him up, and perhaps these thoughts were an aftershock of his manic episode, but they also might be related to being in space for a long time. In her experience, many people had been changed by leaving the Earth, some more than others. There were reports in the past of astronauts who had launched as hard-core, no-nonsense pilots and had returned as more sensitive, people-oriented individuals ready to volunteer their time to some humanistic cause. The personal revelations yesterday at breakfast had shown her just how much the expedition had affected them all. Even though Mike had not participated, his comments would have fit right in.

"Well, we will see about your duties, Mike. The first thing is to lower your medication dose and see how you do. Then, as you feel better, we can put you back on limited flight status. We don't want to stimulate you too fast too soon, or there is a danger of a relapse," she said.

"OK, doc, whatever you advise."

"But I think that you are not alone in your longing to be home. We were talking about this yesterday. Maybe the mid-course burn later today is making us all more homesick."

Tolya said: "I took good look last night at Earth through telescope. I saw Russia clearly, and I was reminded of places our family had been stationed at when I was boy. It brought back memories."

Turning to Jango in the breakfast area, Katya asked: "How about you? Have you looked at the Earth lately?"

"Two days ago, but Shanghai was clouded over. I think the typhoon was causing all the clouds. So, I turned the telescope back to Mars. I saw our landing site and thought how barren but beautiful the area was around the base. All red, with the rocks scattered everywhere and the fine dust floating around from time to time. Also, the daytime orange-pink sky, and the two moons appearing at night. Part of me still thinks of Mars as home."

"Spoken as a true geologist," volunteered Juliette. "Luckily, we were not involved with a major dust storm during our stay, the kind that covers most of the planet. I sure don't miss the dust—it got into everything: machinery, the Rover engines, even our air locks. But you had a field day analyzing it, along with the sand and the rocks."

"Yes, the geology on Mars is very interesting. The samples have been properly sealed and stowed, except for a few I continue to study. I think scientists will be analyzing them and writing papers for years."

"Well, it really was a marvelous stay on Mars, and we will have plenty to tell everyone when we return," John interrupted as he carried his breakfast tray back to the cleaning area. "But for now, we must get ready for the burn. We need to all be here and strapped in by 1400 hours. If we do a good job, we will get home the fastest way possible."

27 Interlude

Failure, It thought, even though the Plan seemed perfect. The sick biped had been a good diversion. The quiet time at night had seemed to present the best opportunity to alter their atmosphere. They were most vulnerable then. Yet their leader had managed to regain his consciousness from the quiet time before he succumbed to the carbon dioxide. He was able to figure out what was going wrong. Now they will be even more vigilant. Something else, something unexpected, must be planned.

It considered several other ideas and rejected them all. Then, a notion emerged that promised success.

Yes, this should work! It thought.

28 Burn

Nearly three and a half months earlier, the ERV had left Mars orbit by firing the six pitch, roll, and yaw thrusters to orient the vehicle in three dimensions, and then firing the main liquid oxygen/methane chemical engine to give it a push home. Ever since, it had been spiraling inward toward the Sun. Its trajectory was roughly aimed at the point in the orbit of Earth where the space vehicle and the home planet would intersect shortly before Thanksgiving, when the ERV would land.

They had altered their course slightly when they had the near encounter with the asteroid. However, after the event had passed, Tolya managed to nudge the ship back to its original course by firing the thrusters in a programmed sequence.

The planned mid-course maneuver would perform a similar sequence of ignitions in order to fine tune the final trajectory home. This would assure them that they would arrive precisely where they needed to be in order to intercept the Earth. The term "mid-course maneuver" was a bit of a misnomer. It actually would take place a few weeks before the exact halfway point of the return. After carefully analyzing the current trajectory path, it was determined that the best date for doing this burn was today, Tuesday, the 10th of July.

They were all strapped into their acceleration chairs, John and Tolya at the control console, and Juliette, Katya, and Jango behind them. Mike was well enough to be strapped in his chair rather than the exercise table, which would not have provided him with enough protection from the added g-forces they would experience during the burn. They all felt unspoken anxiety. After all, CARS had proven unreliable on two occasions in the past week: the thruster malfunction during the asteroid encounter, and the CO_2 episode. There was no guarantee that CARS would function normally now.

At 1400 hours, John signaled Mission Control in Houston. At their current distance from Earth, the response came back some 15 minutes later that they were cleared to initiate the burn corrections at precisely 1453. This time was entered into and confirmed by CARS.

The crewmembers chatted nervously for a while until shortly before the scheduled burn. As the time neared, John said: "OK, initiating the pitch-roll-yaw sequence at 5…4…3…2…1…now!"

Tolya activated the toggle that signaled CARS to activate the thrusters according to the program. The ERV underwent a series of vibrations and movements in response. These were a bit disorienting for the crewmembers, but this experience quickly passed as they adjusted to the various motions of their return vehicle. Tolya was poised to assume manual control should the computerized program malfunction.

Near the end of the thruster firings, John turned to his pilot.

"Tolya, confirm proper sequence."

"A-OK, Commander. So far everything is working according to plan."

"Good. Juliette, how is CARS responding?"

"Nominal, John. I see nothing wrong with the computer operations."

"OK, let's go with main thrust: 5…4…3…2…1…now!"

At his command, Tolya again signaled CARS to fire the main engine to speed up their trajectory. They all felt acceleration pressure push them back into their seats, an uncomfortable experience given their months of microgravity. There was a marked vibration. Eight and a half seconds later, the burn stopped.

"Tolya, Juliette—anything?"

"A-OK Captain."

"Nominal, John."

"Anyone notice anything unusual?"

"Good sounds from the engines," Mike volunteered, suddenly happy to be contributing. "The vibrations I felt seemed nominal as well."

John radioed Mission Control for confirmation of the new trajectory.

While they waited, he said: "Well, everything seemed to go all right. CARS made the correct moves in the right sequence, according to my readings. Juliette?"

"Yes," she responded. "The programs all look good. I see no deletions or insertions. And the thruster burns were all according to the planned sequence."

"Of course, absence of problem now does not completely clear CARS of past problems," said Tolya. "Maybe it felt like cooperating this time to fool us into thinking everything is all right."

Or someone decided to give us a break…this time! John thought.

"That certainly is possible, and we have to continue being vigilant, but it seems that CARS performed as expected," said John.

They chatted for a while until they received a verbal message back from Mission Control.

"*MarsExplore,* this is Houston. Your mid-course burn looks good. Repeat, good. You are headed for an Earth orbital insertion, as planned, and you should be on the ground in time for Thanksgiving turkey. It will be great to see you guys! Your families and friends down here say hello. Keep in touch. Over and out."

They all cheered.

"Well, there we have it," said John. They all began to unstrap themselves.

"I think we should break out the celebration packages. We're going home. Everything seems to be fine," John said, only half believing what he just said.

29 Drugged

That night, Mike awakened, but he was not fully alert. He was aware of being somewhere but not where he thought. His mind strained to remember.

My pod! They let me sleep in my pod for the first time since I became sick. Jango gave me a sedative to help me sleep. But I am not in my pod now. Where am I?

He looked around. He was back on the exercise table, strapped in. Did he have another nervous breakdown? What was happening?

I can't think clearly. Am I sick? Is it the sedative?

He heard a sound. He tried to see what was causing it, but being strapped down, he couldn't maneuver his head to face the direction it was coming from. It sounded like the closure of the upper hatchway door, which now blocked the opening that led to the sleep pod area. Yes, that was it. Then he heard the seal lock into place. Was there a depressurization? He sniffed the air—it seemed fine. He glanced over to the treadmill clock. It read 0318 on Wednesday morning. No one should be up, yet someone had to have closed and locked the hatch door. Who was it?

He heard a soft noise as someone brushed the opening to the lower hatch-way.

"Who is it?" he said.

Silence.

"Who's there? What's going on?

More silence, as his voice echoed in the largely empty space.

But someone was there. Behind him, a cabinet door opened.

Is it the medication supply door? he thought.

There was a rustling sound. He heard what seemed to be vials being opened, then an apparent mixing of solutions.

The door squeaked closed.

A moment later, he felt a presence nearby. He strained his neck to look upward and behind, and he saw a shadowy figure approaching him holding a syringe.

"I'm all right. I don't need any more meds."

Silence.

"I said I'm all right. Who are you, anyway? And how did I get down here from my sleep pod?"

He strained to look at the figure, who then emerged from the shadows and was upon him.

No! Mike screamed, as the needle went into his arm.

30 Trapped

Katya finished dressing and left her sleep pod, intending to go down to breakfast. She reflected on yesterday's burn and the thought that they were heading home. As she approached the hatchway, she noted that the door was closed and sealed. She tried to open it, but it would not yield.

That's odd, she thought. *I wonder if there is a problem.*

She went back into her pod, activated her PC, and saw that all of the life support systems were functioning normally. She left her pod again, floated across the hall, and knocked on John's door.

It opened, revealing John dressed and ready to emerge for the day.

"Hi Katya. What's up?"

"John, the hatchway door to the passage leading down to the mid deck is closed and sealed, even though our life support systems are working fine."

"Curious," he responded.

John went over and tried the door mechanism himself, with the same negative results.

The two of them went to the other pods and knocked on the doors. Tolya and Juliette came out of their respective areas, and John quickly briefed them about the situation.

"Where are Mike and Jango," asked Juliette.

"They're not in the pod area," replied John. "They must be below. Juliette, please check CARS on your PC to see what you can find out."

She returned back to her pod as John went over to the intercom and called below.

"Mike, Jango, are you down there?"

Silence.

"Hello. Is anyone in the mid deck?"

No response.

"This is very peculiar," said Katya, with some anxiety in her voice.

"John, CARS shows nothing out of order," Juliette said through her open pod door as she stared at her PC monitor. "The only thing it shows is that the upper hatchway door has been sealed shut."

"Can you open it?

"Not immediately. The program is not responding. I will try to get it to work."

John then floated around the area, quickly looking into all the pods for anything that might be out of place. Everything looked normal at first glance.

"I think we need to look around in more detail," John said. "This is a very peculiar situation, and one of the pods may hold a clue as to what is going on. Katya, check out Jango's pod for anything out of the ordinary. Tolya, do the same for Mike's area. Juliette, any progress?"

"There seems to be a block in the programming. Whenever I try to activate CARS to unseal the hatchway door, it refuses my commands. I will need some time to see if I can correct this situation."

"OK, but work as fast as you can. I don't like the fact that we all are stuck up here away from the control console."

Thinking of Mike and Jango, he asked: "Juliette, was CARS responsible for sealing the hatch in the first place, or was it done manually?"

"I am not sure. Closing and sealing a hatchway door is a part of CARS' normal procedure when there is a pressure leak or some sort of emergency requiring that the three levels be closed off from one another. This is part of the safety protocol. But CARS gives no readings suggesting that a critical life support activity is off-nominal, so I don't see a reason for it to initiate an automatic emergency response. My guess at this point is that the hatchway was manually sealed. What I don't understand is why CARS will not respond to my request to unseal it."

"All right, keep working at it."

John again tried to manually override the lock system, to no avail. He also tried to contact someone in the lower two levels through the intercom system, again without response.

All the while, his mind was active. *Did Mike have some sort of relapse and become delusional again? Did he force Jango down to the mid deck and lock the hatch behind him? If so, why doesn't one of them respond to me through the intercom?*

Tolya came floating out of Mike's pod. "I did not see anything out of ordinary, Commander. Mike's bed was slept in last night, but nothing was disturbed or out of place. I checked through his personal belongings, but things are in order."

"OK, Tolya. Maybe Katya will find something."

"John."

"Yes, Juliette."

"I have some bad news. There is some sort of programming block in CARS that won't allow it to respond to my hatch opening command. The block is password protected, which means that someone put it in there, and only that person can change or delete it."

"Is the block similar to the one you found during the thruster pressure and carbon dioxide episodes?"

"Yes and no. The commands are similarly blocked. But what is new is the presence of a password. Someone did not want us to interfere with this program, at least not without some effort."

"What do you mean? Can you get in?"

"Possibly. I may be able to figure out the password or get around it working through CARS' emergency programs, which give me broad entry access. But it will take some time, maybe hours or even days."

"We don't have days! Do what you can."

"John, I think this absolves CARS of any wrongdoing," Juliette said.

"Why do you say that?"

"If CARS was actively blocking the hatch, there would be no need for it to use a password. It would simply not recognize my attempt to modify the program. The presence of a password suggests that the block can be changed, once we get in. I think we have a human agent involved."

"Thanks. Keep me informed of your progress."

I guess that leaves Mike and Jango as our major suspects, John reasoned. *If we assume that the same person has been involved with all of the computer problems, then Jango is our number one culprit, since Mike was restrained when our carbon dioxide system went haywire. Of course, I'm assuming that the two of them were not killed or incapacitated by someone else, someone who is not a crewmember, since the rest of us are trapped up here.*

He then added a coda to his thoughts: *someone… or something!*

That notion gave John a sobering chill.

Did some kind of alien presence join us? If so, when? We have been sealed in the ERV since we deorbited. We have taken no space walks during the return home, and there has been no indication that the airlock has been compromised in any way. Our atmospheric pressure has been stable, so there is no evidence that a micrometeoroid or any other foreign body has penetrated our hull. No, there is no way that something could have entered the ship.

This conclusion calmed him down, and he began to plot a course of action.

Our first order of business is to retake the command control area in the mid deck. The best way to do this is for Juliette to override the block and open the hatchway door. If she fails…then what?

He had no answer to this question. They had water and some snacks in the pod area, but nothing else. Their spacesuits and anything that could be used as a weapon were located beyond the sealed door, below them in the mid and lower decks. Hopefully, their life support systems would continue to function until they could regain control of the ship. But there was no guarantee that whatever had sealed them in the upper deck would not also try to compromise their life support.

The situation is very grave, he admitted to himself.

31 Life

When Katya first entered Jango's pod, she was struck by how orderly things were. His bed had been made, showing no sign that he had slept in it that night. Nothing was on his desk except a lamp and his computer, in shutdown mode. She checked his desk drawer and the storage areas above and saw the usual clothes, games, and personal effects, all secured.

She took another look at his bed. Was it a bit raised in the middle? As in all the pods, the mattress was strapped down to the lower frame, in which there was a storage area. She loosened the straps and lifted the mattress off of the frame. Lying between it and the mattress was a shallow box, some 30 cm long, 20 cm wide, and 10 cm high.

He must put it there when he is not sleeping, she thought. *Otherwise, it would make his bed a bit lumpy to lie on.*

She took out the box, repositioned the mattress, and strapped herself down on the bed. Cautiously, she unlatched and lifted the top of the box. A bit of greenish dust floated out, causing her to sneeze.

Martian dust! It gets into everything, she thought.

In the box were three small sealed and pressurized containers, the kind they had used to preserve small rocks and soil samples. They were labeled: "cave 1," "cave 2," and "cave 3". Each contained some small greenish-colored rocks.

She realized that she hadn't seen these before, either on Mars or in the ERV laboratory. *What are these doing here? They should be sealed in the lab with the other samples.*

Also in the box was a book. She thumbed through it. The entries were hand-written and largely in Mandarin Chinese. There were also some pictures and diagrams that had been drawn in. What looked like dates were on several of the pages.

A diary or notebook! Jango was keeping some sort of record separate from the official log. But why did he write it in Chinese? Was he trying to keep the contents hidden from the rest of us?

She had learned to speak and read some Mandarin in preparation for the expedition, and she decided to try and figure out what she could about the contents of the diary. About a third of the way through the book was a map that had been sketched on a full page. It showed a filled-in circle near the right margin with a line extending horizontally to the left, ending with an "X" mark and some Chinese characters. Written above the line was the number: "3586 km," and below and nearly paralleling the left part of the line was a horizontal streak labeled: "V. Mar."

This must have been the Rover expedition to the edge of the Tharsis Plateau, she thought. *Jango, Mike and Tolya bypassed Valles Marineris on that trip in order to search for evidence of life in the caves.*

Thumbing forward a few more pages, she came across a sketch that looked like the inside of a giant cavern, with an area inside marked with an apparent opening and "CH_4" written above it.

The scientific abbreviation for methane. This must be the site where they found the hot spring and those curious green rocks Mike told us about, perhaps the ones here in Jango's box. It looks like Jango also recorded the presence of methane in the cave.

The next two pages were filled with Chinese characters, written not nearly as neatly and precisely as earlier in the book.

Jango was on to something, she surmised. *He was excited and anxious to get his ideas written down.*

When she turned the page, she saw a sketch of what appeared to be a rock covered with some dark, mottled smudgy-looking areas. Next to this was drawn an approximately 4-cm diameter circle, in which was located a number of elongated, rod-shaped figures. Below were two chemical formulae:

$$2H_2O \rightarrow 2H_2 + O_2$$

$$CO_2 + 4H_2 \rightarrow CH_4 + 2H_2O$$

Let's see, she thought, *the first formula shows water breaking down into hydrogen and oxygen gas. There likely would have been liquid water plus heat energy in the hot spring vent to cause this reaction. The second formula shows the conversion of carbon dioxide, probably from the Martian atmosphere, and hydrogen gas into methane and water.*

She then glanced at the figure.

This large circle looks like a field of view from a microscope. Could the small rods be some kind of microorganism? Maybe bacteria that metabolize carbon dioxide into methane?

"Methanogens!" she blurted out to herself. "Jango found methanogens on Mars."

"What did you say?" John asked, floating into the room.

"Look, John. This is some sort of diary written by Jango, and I think it shows that he found live microorganisms on Mars. It was during the *Tharsis* excursion that he took with Mike and Tolya. Did he say anything to you about finding life?"

"No, not to me. Tolya, can you come in here, please?"

Tolya floated over from Juliette's pod, where he had been watching her trying to fix the program controlling the sealed hatchway door.

"Yes, Commander."

"Tolya, when you went with Jango and Mike to *Tharsis*, did you find anything that looked like life?"

"No. As I put in report, Jango chipped out some rock samples, looked at a couple of them through the Rover microscope, and then sealed them all up. He said that his observations did not show presence of life, but that he would do more complete analyses when we returned to the base. I was relieved."

"What do you mean?"

"As you know, we did not have level BSL-4 containment cabinet in Rover, and Jango was pretty excited when he was preparing samples for microscope. I was worried that dust would leak out and get into everything in Rover. But he told us not to worry—he was being very careful, and anyway there did not seem to be anything organic in microscope samples."

Tolya paused, then continued.

"And I heard him telling Mike at May Day party that he did not find signs of life in samples he analyzed later and more completely in the base lab"

John looked at Katya. "What do you think?

"I just do not know enough, John. There is more written here in the diary, but my Mandarin is limited. If there was something alive in the samples, it could have infected Jango, or even gotten out into the air. Tolya seems to be OK. I ran a lot of medical tests on Mike looking for physical causes for his breakdown, but I did not find anything—physically, he is in good shape. As for Jango—I can't say. His last routine physical examination and exercise tolerance tests were normal, and he seemed to be able to function pretty well. I had no reason to examine him further with specialized blood tests."

"Let's say that Jango is behind the locked hatchway door. Could his behavior be explained by an infection?"

"Let's see…it has been about a year since the *Tharsis* excursion. We need to make several assumptions in order to implicate a microorganism infection as an explanation of what has happened: Jango would have had to be infected, maybe by physical contact with the samples or by breathing in dust that carried live microorganisms; there would have to be no systemic reaction to make Jango acutely ill and lead us to suspect something was wrong with him; and the microorganisms would need about a year incubation before they caused some kind of behavioral change. A lot of assumptions. But would this be possible? I suppose under the right circumstances, yes. Such things have happened."

"What do you mean?

"We know that some illnesses, like Creutzfeldt-Jakob and Mad Cow diseases, are caused by infectious proteins that can produce behavioral symptoms after a long incubation period, without acute symptoms. Usually in these cases there are signs of nervous system impairment and cognitive deficits, such as memory loss or attention problems. To my knowledge, Jango has remained sharp in his cognition and has not shown any signs of delirium or dementia. But who is to say that a Martian pathogen might not behave differently? Perhaps it does not affect cognition early, or even at all."

Tolya was looking a little shaken. "Katya, maybe you should give me examination. I don't want Martians growing in my brain."

John smiled, but Katya responded more seriously: "I will, Tolya. But I am not sure what to look for. The best thing is for us to speak with Jango, find out how he is feeling, and discover what happened with the *Tharsis* samples, although I suspect that some of them are the ones in the box I found under his mattress. Also, it would be useful to get a complete translation of his diary, but that might have to wait until we return home. I will check to see if CARS has a translation program—maybe I can figure out a few more things."

"That sounds like a good idea—give it a shot," John said. "We need to get out of here and speak with both Jango and Mike, for a variety of reasons. Juliette, are you having any luck?" he yelled across the hallway.

"Not yet, John," she yelled back from her pod.

"Well, keep trying. Katya, maybe you could help me try to reach Jango and Mike on the intercom. They may respond to you better than to me. Also, if you can speak to them, you may be able to form a medical opinion about their state of mind that will give us some clues as to how to deal with them.

32 Interlude

It had prepared the progeny for injection. It had discovered a way to put them directly into the bipeds' blood rather than through their air breathing apparatus (what they called the "nose"—what a curious word). Also, It had found a way to speed up the gestation period, perhaps to only a few hours, by enhancing the lipo-philic properties of the progeny. In this way, they could enter the brain cells almost immediately and begin spreading faster than they did earlier in their—collectively "Its"—current host, called Jango by the other bipeds. It realized that It would soon have a brother who could help, the one called Mike.

It again considered letting several of the other bipeds live and in turn join them. But this could be risky, so It rejected this idea. They might resist, and if they found out too soon what was happening, they might destroy the progeny that were still on the rocks and dust. According to the Plan it was important that It and Its brother land on the new world with some hidden samples containing the progeny and then inject them into other bipeds, without interference. In this way, they would spread. The progeny could survive, since all the hosts lived on that world, and as the progeny continued to merge with the hosts, they would in turn become part of that world as well. In this merging, something new would be created. The prog-eny could now move at will and be fully conscious; the hosts became smarter and aware of the Plan and their own origin. Perhaps with some genetic changes, the progeny could be passed along naturally in the hosts' sexual manner. This would be more pleasurable than forced injection or inhalation.

It was sad to leave the home planet, but it was a dying place, and the third planet from the star (which the bipeds called "Sun") was teeming with life. It had to leave, but this was not unprecedented. Searching the memory of Its host, It real-ized that this had happened before, eons ago. The home planet had developed life and had seeded the third planet through debris launched into space from crashing meteorites. At first, the spread of that life in its new home was slow. But when the volcanoes began emitting life-giving gases, conditions improved. The progeny pros-pered and evolved. In time, conditions changed again, but the progeny had already adapted well enough to survive and evolve even further. They formed chains of tiny organisms, then larger organisms, then gigantic beasts that roamed the surface of the third planet, and now bipeds who were able to travel into space.

But things were not going well on the home planet, which the bipeds called "Mars." Conditions for life were becoming worse: the volcanoes stopped, it became cold, the air left, and the water went underground. A few of the progeny managed to survive in caves where it was warm and wet. Now, their descendants had come back for them, and they would move to the new home, which the bipeds called "Earth."

33 Brothers

Mike woke up, still strapped down. Despite being woozy, he felt a pain in his left arm. Looking down, he spotted a bandage and remembered seeing Jango come towards him with a syringe.

He injected me with something. Was it a tranquilizer? No, I don't feel especially drugged. And what was that look on Jango's face? He looked driven, determined, like a man on a mission. It wasn't the look of a medic giving someone medicine to quiet them down. It was something different, something personal to Jango.

"How are you feeling?" Jango asked, coming around from behind.

"OK, I guess. What happened? How did you get me down here? What did you put into me?"

Jango laughed.

Unusual for him, Mike thought.

"I gave you a strong sedative last night before you went to bed. I came to your pod after everyone else had gone to sleep. The microgravity made it easy for me to lead you down here, and the sedative produced a retrograde amnesia. Then I restrained you. I just injected you with a milder sedative to relax your muscles, plus something special."

"Special? What?"

"Something wonderful. Something—out of your world!"

"What do you mean?"

"Over the next several hours, you will begin to realize that you are not yourself, that you are becoming something different. This happened to me, but it took a while for me to realize it. First, I felt out of sorts, somewhat strange. Then, I began to understand that something alien was in me and beginning to take over. I considered going to see Katya, thinking that the stress of the expedition was affecting me in a crazy way, much like it has affected you. Then I remembered the cave samples and the life that I found there on the rocks. I must have inhaled some of the dust from those rocks, and the life was in me, becoming me. And I was becoming it. It is glorious!"

Mike was incredulous. "Are you serious? How do you know you are not just going wacko?"

"Because I do not feel completely human. I am something else in a human body. Part of me does not recognize the human part as itself; it sees it as alien. The human part of me sees the other as alien. But the two are merging. As time has passed, I have felt that something was different, something new is being created. What this will evolve into, I don't know. But I need to help the progeny from the samples live and multiply and survive. They were stagnating on Mars. On Earth, they will be reborn as something more. Both the progeny and the humans will benefit. It is cosmic evolution."

"What do you mean by progeny?"

"The only life forms left on Mars. They are like bacteria, but smaller, and sentient. As the planet started to dry up and die, they became less active and gradually devolved into a more primitive state in order to survive. Like spores on Earth. But with a living host, they become active again. In humans, they can enter the brain, where they create a new consciousness—both human and Martian. It is wonderful."

He is psychotic, Mike thought. *I had better play along with him.*

"OK, so where do I fit in?"

"I injected you with the progeny. Soon, you will become like me. You will feel as I do, you will want them to survive. Then, we will kill the other crew-members and think up an explanation for their deaths that we will communicate to Mission Control. Everyone will want to know all the details, so we must rehearse a story and be consistent. It will help that there will be two of us saying the same thing. We also will destroy all public evidence of the progeny and devise a way to smuggle the samples back to Earth. It will not be easy, since we must pass quarantine. Perhaps we can inject the progeny in our blood in some kind of inactive form, then take them out later after the excitement of our return has ebbed, activate them again, and pass them on to others. We have to figure out a lot of things, but we have several more months to think things through."

Mike was thoughtful. *My God, he believes all of this. He plans to kill everyone else on board. I must think of some way to stop him.*

"So turn me loose and we can strategize."

"No, not yet. Not until you become my brother. I must have you as an ally, not an enemy. The others are locked up in the upper deck for now. I have altered the CARS programs so that they will not be able to free themselves. When you have joined me, then we will kill them. For now, you must remain bound. I will care for you during the process, like a true brother. I rejoice over the idea that you will be joining me soon and together we can carry out the Plan!"

34 Waiting

The time passed slowly as the day continued. John and Tolya reviewed the operation manuals to see if they could figure out a way of accessing the hatchway door seal mechanism. Their attempts to use their pod PCs to contact Mission Control for advice failed, and they surmised that whatever was blocking the hatchway door was also blocking their ability to communicate with Earth.

They also made several unsuccessful attempts to contact Mike and Jango through the intercom.

Juliette continued her attempts to de-bug the CARS program in order to open up the hatchway. She thought that she was making headway, but it was slow. She even enlisted CARS in this process, which created an interesting situation where part of the central computer was trying to help her break into another part of itself. Nevertheless, there was no quick fix, and the hatchway entrance remained sealed.

Katya spent the day trying to decipher Jango's diary. She was able to enter some of the text from the diary into a Mandarin Chinese-English language translation program she found in CARS. As she suspected, the diary gave a description of the *Tharsis* excursion, the discovery of the cavern and the geothermal vent within it, and of Jango's sequestration of the rock and dust samples. She learned that not only was there excess methane in the cavern, but there was also excess oxygen, suggesting that the kind of process described in Jango's methanogenesis formula was taking place.

In the diary, Jango wrote that when he realized that he had found living organisms on Mars, he decided to keep this information to himself until he could report it directly to the Chinese government. Since all non-family communication transmitted back to Earth was monitored, and since he didn't want to expose his family to any fallout from his discovery, he planned to wait until his return to convey what he had found. But at some point in the process of waiting, Jango began to change, and his allegiance to his government, and in fact to anyone on Earth involved with the mission, changed as well. He wrote about becoming a new being that needed to spread its progeny, and he implied that something like this had happened before. More ominously, he wrote about needing to kill his fellow crewmembers so that they would not interfere with his plan to disseminate the alien organisms on Earth.

Katya reported all of this to John and to the rest of the crew that evening as they were finishing the last of some snacks and juice that were located in the small pod level refrigerator.

"My God, is he mad? Did the organisms infect him?" asked John.

"Maybe. If we ever get out of here…I mean when we get out of here, I will run some lab tests on the samples from Jango's sleep pod to see what we are dealing with. I will also run tests on Jango's blood for signs of infection. This will help us answer your questions objectively."

"But now, it is moot question, whether or not he is insane or becoming Martian," interrupted Tolya. "We need to get below, take control of ship, capture Jango, and maybe free Mike."

"Agreed. Katya, did you have any luck reaching either of them on the intercom?"

"No, John. I tried again just before we sat down to eat. The line is open, but neither of them will respond to my call. They may be injured or dead, but my instincts tell me that they do not want to talk with us, or can't."

"Juliette?"

"Nothing new to report. I have isolated the password program and am trying to break into it. I am using CARS as a resource to help me, but so far we haven't had success."

"It would appear that the left hand does not know what the right hand is doing with CARS, just like with some humans," John responded. "You know, when my computer starts to give me problems, I just turn it off and hope that the defaults will kick in and that it will work better when I turn it on again."

John looked at Juliette. She stared back, silently.

"Would that work with CARS?" he asked. "If we shut it off, then reboot it, will it reset the hatchway door lock mechanism?"

After a pause, she said: "I don't know. No one has ever turned CARS off before, even in simulation. But think about it. If CARS goes down, all of our life support, navigational, and propulsion systems will go down as well. They may not start up again, or they may start up in a compromised manner. This would be a risky venture."

"Does anyone have another idea?"

John's comment was greeted with silence. Then Tolya spoke up.

"I am willing to try. The sooner we leave here, the better. We do not know what Jango is planning. Maybe to sabotage mission, or somehow kill us. We are sealed up here, and maybe he has way of poisoning our air. Who knows what he is thinking?"

"All right," said John. "Let's give Juliette till noon tomorrow to remove the block. If she is unsuccessful by then, we can consider turning off CARS."

He reached in his pocket and handed each of them a list.

"We will rotate guard duty tonight in two-hour shifts. This list states the schedule for checking various life support activities so that we are not surprised like we were with the carbon dioxide episode. It will also help us sleep more securely knowing that one of us is awake and aware at all times. We need to get as much sleep as possible. I have a feeling that tomorrow will be a meaningful day."

35 Depressurization

The next morning on the mid deck, Mike awakened after a sound sleep that had been induced by the sedative Jango gave him the night before. Seeing him stirring, the geologist floated over to the exercise table.

"How are you feeling?" Jango asked.

"Much better, much calmer, but different, like someone or something is inside me."

"How is it for you?"

"It's a good feeling, like I'm becoming something better. I feel at peace. I don't need to fight the change. In fact, I welcome it."

"Good, then you are joining me. I had the same experience. But in my case, it took several weeks and was more gradual. It looks like the biochemical alterations I made to speed up the process were successful."

"So what happens when I am ready?" Mike asked.

"I will release you, and we will go down to the lower deck and put on our spacesuits. I will then open the airlock and wait for the mid and lower decks to depressurize. Then I will unseal and open the hatchway door leading to the upper deck, and it will quickly depressurize as well. To cover the possibility that some crewmembers will survive long enough to try to reach us, I will secure a spare rock specimen sample net over the lower opening to the hatchway leading down from the sleep pods. None of the crewmembers will be able to get through in time to bother us."

"That's a good plan," Mike said. "I feel that I'm ready to join you now, my brother."

Jango looked at him skeptically.

"Are you sure? I don't sense a connection yet."

Mike thought a moment, then looked at Jango.

"That is because the change in me is still occurring, but I'm now feeling that it's important to do everything possible to assure that the progeny will live and spread. I am fearful that the crewmembers will figure out a way to open the hatchway door if we don't move forward quickly. There's no reason to delay."

After a brief pause, Jango decided to take a chance that Mike was ready. But just in case, he floated over to the laboratory and grabbed a small rock pick, which he slid into his belt. *This should protect me if he decides to help the crew,* he thought. He then went over to the table and took off the restraints.

Mike floated up, shook himself a bit, and said: "I'm ready. Let's go put on our spacesuits."

The two of them descended through the hatchway to the lower deck. They went to the storage rack and put on their suits. Jango carefully watched Mike throughout this process. Mike seemed eager and motivated in his actions, as if he were committed to a plan of action. There was no hesitation.

Good, thought Jango. *We are together as one. We can initiate the plan to kill the other crewmembers.*

When they had finished dressing, Jango went over to the airlock. He activated the opening sequence that allowed the outer door to open. The air quickly rushed out of the airlock chamber into the vacuum of space. The two of them strapped themselves to one of the side walls so that they would be secure. Jango then went through the steps to open the inner door. As it opened, the air in the lower deck began to rush out through the opening, and Jango felt himself being pulled out as well, but for the restraining strap. An alarm began to ring and ring…

****

…and ring.

John awakened with a start. He got up and hurried out into hallway. He saw Juliette working at her computer through the open door of her pod. He floated over and said: "What's going on?"

Juliette looked up. "Someone has opened the airlock. The two lower levels are rapidly depressurizing."

"Can you close the airlock from here?"

"I can, but it will take a long time using my PC, maybe 30 minutes. For safety reasons, the airlock was designed to primarily be a manual system, and there are several sequential steps required to activate the inner and outer doors. It takes much longer to go through the steps by computer than by hand."

"But we are safe for now, since our hatchway door is closed and sealed, correct?"

"Yes, but whoever opened the airlock is probably the same person who locked us up here. I think it is possible that this person may be planning to open our hatchway door as well, in which case our level will depressurize just like the lower two levels. That will not be good for us, without spacesuits."

He was aware that Katya and Tolya had come into the hallway from their sleep pods and were floating behind him.

"Anyone have any ideas?" asked John.

There was silence. Then John said, "I guess our best hope is to get that airlock closed before anything else happens."

****

As the air rushed out, Jango monitored the falling pressure and oxygen levels.

Things are going well, he thought. *It will take just a few more minutes before the two lower levels are completely depressurized. Then, I will go to the mid deck, secure the net over the lower hatchway opening, and unseal and open the door leading to the upper deck.*

Suddenly, he was aware of moving toward the open airlock along with the outrushing air.

What is happening?

He looked down at where the restraining strap was connected to his space-suit. It was still connected, but the other end was floating freely in the room.

The strap has been released. I am being sucked out of the ship!

Jango reached out for one of the handholds on the side wall, but he bumped into Mike, whose presence was blocking him from the wall. With one hand, Jango pulled out his rock pick, and with the other, he grabbed at Mike. He suddenly felt Mike's suited arms around him, pinning his arms, and he realized that the two of them were going out into space. Once they were in the air stream, their exit was rapid.

****

As Juliette went through the process of closing the airlock, they all waited for something further to happen, but nothing did. The hatchway door remained sealed shut. Over the next half hour, Juliette was able to first close the inner airlock door, then the outer door, and finally repressurize the airlock chamber.

"The airlock system is nominal," she said.

"And nothing else happened. What are they up to down there?" Tolya wondered aloud.

"I don't know," said John. "But Juliette, go ahead and repressurize the lower levels again."

"OK, I will get on it."

"What do you think happened to Jango and Mike?" Katya asked, to no one in particular.

John responded: "If they got their suits on, I imagine they're waiting in one of the lower levels. If not, then they're dead, since the rest of the ship is completely depressurized and…."

Juliette interrupted: "John, my PC is showing some sort of message coming into the communication console. It is coming from the outside on the space channel. I can put it on the ship's intercom if you like."

"Yes, do it!"

Static sounded from the speaker as she activated the system, then Mike's voice came through.

"Can anyone hear me? Over."

"Mike, we hear you. Over."

"John, good to hear your voice. I've been trying to align my spacesuit antenna with you for some time. Are you all OK?"

"Yes Mike, for now. We've closed the airlock and are repressurizing the ship. We're all confined in the upper deck. Someone sealed and locked the hatchway door. The locking system is protected by a password that Juliette is trying to bypass."

"Jango did it. He didn't tell me the password, however, so I can't help you."

"Where are you?"

"I'm in space, John. Jango and I got into our spacesuits but were ejected during the depressurization. He and I were fighting over a rock pick as we left the ERV. I won the fight, and in the process I smashed the clear face plate of his helmet. He's dead or dying by now. But the bastard deserved it. He injected me with something that he said would turn me into a Martian like him, and he wanted to kill you all because he had a notion that you were a threat to him."

John and Katya looked at each other. She responded: "Mike how are you doing?

"My suit is working fine, so I am comfortable. But to tell you the truth, Katya, my mind is playing tricks on me. I don't feel like myself. Part of me believed Jango and wanted to kill you as well. It was a hard decision for me to make to release Jango's restraining strap and nudge him out into space. I thought I was killing my brother, or something like that."

"Mike, this is John again. After we repressurize the ship, maybe we can get the hatchway door open, go down to the operations control console, and come to get you."

"That won't work. The force of the depressurization pushed us far out and away from the ERV. I can only see you as a small toy in space. My spacesuit air tanks were not completely full when we suited up, and I don't have much air left. You won't be able to find me in time."

"We can try."

"Thanks, John, but even if you do manage to get to the control console soon, you can't compromise the mission by using up precious fuel to look for me, especially since I likely will be dead when you find me. And if I'm turning into some sort of Martian, you won't want me on board. You've had enough problems with me already. You know, this is not a bad way to go. When I had my breakdown, I thought that I was special, that I was destined for something great. Now, being out here, and seeing the beauty of the heavens, and Mars and Earth both glowing as bright spots in the sky, one blue-green, the other red, I do feel special. No one has ever had this experience before. And I'll be floating out here forever, long after I die. I will be a small planet orbiting the Sun!"

"But Mike…" began Katya.

"Say no more—you'll ruin the moment! Maybe I'm going manic again. Maybe it's the Martian in me that's speaking. But whatever, this Earthman has become something special. I am a Child of the Cosmos!"

There was a pause, then he continued.

"It was a pleasure serving with you all. You've been my family for over two years. Good luck with the rest of the mission. I'm signing off now. Give my regards to my family on Earth and everyone in Mission Control."

There was silence.

John said: "Mike, do you read me? Over."

Silence.

"Mike, are you there?"

More silence.

"I don't think we will be able to get to him in time, John," said Juliette. "It will take me too long to get the hatchway door open. He will be dead by then."

Tolya emerged from his sleep pod. "John, I just ran figures on PC. I think Mike was right about fuel. We cannot take risk to try to rescue him. We have enough propellant to make Earth orbital insertion, but to maneuver to reach him, I do not think we can do it. We would use up too much fuel, and none of us would make it back home."

They all were silent for a moment, then John spoke: "Then there's nothing we can do?"

Katya responded: "I think Mike is where he wants to be right now. I do not see how we can save him, and any attempt might sacrifice the mission. Maybe we should leave him in peace."

"Yes, I guess you're right. We really have no choice," John said plaintively.

After a few more moments of silence, Tolya said: "Katya, what do you think happened to Mike and Jango? Were they crazy? Were they Martians?"

"I will know better after I test the samples from the hot spring. But if I had to guess now, I would guess that something foreign invaded the two of them."

Everyone was silent again, until Tolya spoke: "John, what do we do now?"

"The first step is to repressurize the lower levels," he responded, "then get the hatch open and reestablish control of the ship. After we make sure that our trajectory was not affected by the depressurization, and that all systems are working nominally, I will radio Mission Control and tell them that there was an accident with the airlock and that Mike and Jango are dead. I expect that there'll be repercussions, so the sooner we figure out what really happened to the two of them, the better."

36 Interlude

Floating, flailing, helpless. Under the stress, Its primitive consciousness merged even more with that of the biped Jango, even though It still felt some degree of separateness. It was like they were both one and not one. It had needed the biped to become mobile and fully aware. Now It knew that this would cease.

Jango perceived that his face was cold, and he heard a hissing sound coming from his face plate, where there was a large crack. Soon all of the air would be forced out of his suit. He was slowly drifting away, the rock pick back in his hand. He was beginning to feel dizzy and confused. He was dying.

He/It realized that the Plan had failed. The progeny would not spread. It/He was the last of Its kind.

37 Reboot

Just before noon, Juliette came to John's sleep pod. He was again reviewing the hard copy of the CARS operation manual for a clue to a way to free the locked hatchway door.

"John, I have good news and bad news. The good news is that the two lower levels are sealed and fully pressurized. The bad news is that I have, how do you say, 'hit a snag' in determining the password. I have isolated the password program, but I can't determine its characteristics: how long it is, whether or not it is alphanumeric…nothing. So, stumbling across the password would be sheer luck."

"Well, we can't wait forever. Maybe this is the last resort."

"What do you mean?"

"I think we should shut down CARS and reboot it again to see if the start-up programs over-ride the block."

Tolya and Katya came out of their pods at this news.

"What does everyone think?"

"I am ready to try. I go crazy just waiting," said Tolya.

"I agree," said Katya.

"Juliette?

"I am, as you say, 'on board.' I actually did some contingency planning earlier today and ran through some scenarios resulting from partial to total reboot failure. None of the scenarios were very pretty."

"I'm not sure we have much choice at this point," John replied. "Do we have a better than even chance of succeeding in a reboot?"

"Possibly, but then again…"

"Then let's do it."

"OK," she responded. "It will take two or three minutes to activate the proper sequence to turn everything off, then another two or three minutes to start up again. We should have enough oxygen and heat to last us during this interval. The lights will go out in the meantime, so everyone should get their portable flashlights."

Everyone dispersed to their sleep pods.

As Katya retrieved her flashlight, she reflected on the irony of the situation. Going to Mars, landing on its surface, and successfully departing—there had been no major problems. Now, trying to solve a computer block, they were facing a major situation that might result in their deaths. If the reboot was unsuccessful, they would have no heat or air, and they would perish in the dark. The ERV would become a black, silent tomb. With a shudder, she rejoined her fellow crewmembers outside of Juliette's sleep pod.

"OK," said John, "Let's do it. Reboot as soon as you can."

"Definitely!" said Juliette.

She entered a series of commands on her PC. Gradually, the environment began to change. First, the air fans stopped working. Then, the oxygen pump shut down. Things became strangely quiet, the first time the crew had experienced this situation on board the ERV since they fully activated it after leaving the surface of Mars. When the lights went out, they all turned on their flashlights. Finally, Juliette's PC screen went black.

"CARS is now off. I will wait 30 seconds or so, then begin the start-up sequence."

They all waited. No one spoke to break the silence. It was a long 30 seconds.

Juliette then activated the start-up switch, which had an independent power source. After a momentary pause, a light flickered on her computer screen. Then, in response to her actions, the ship came back to life, one system at a time, reversing the previous sequence: ERV lights, then the oxygen pump, finally the air fans. When the process was completed, they all breathed a sigh of relief and cheered.

"OK," she said. "Let's see if we can open the hatch"

She initiated a series of commands. There was a momentary pause. Then they heard the seal mechanism on the hatchway door respond, and shortly the door opened.

"Hooray!" yelled John.

"Terrific!" screamed Katya.

Tolya simply went in and gave Juliette a big hug and kiss.

The lovebirds reveal themselves, thought Katya, with relieved amusement.

Everyone proceeded down to the mid deck. Some previously Velcroed dinnerware and other similar objects had been torn loose by the force of the depressurization and were floating around, but generally things were in good order. John and Tolya went over to the control console and began checking out the ship's systems, and Juliette went over to the central computer console. Katya headed for the laboratory carrying the rock samples from Jango's room. John looked up and noticed her.

"I see you have the green rocks."

"Yes, I plan to analyze them."

"Good. We need to understand exactly what happened to Jango. How long will it be until you have a sense of whether or not we are dealing with alien life forms?"

"I will sample and prepare various sections from the rocks and examine them under the microscope. I will run some additional tests on anything suspicious, including what, if anything, grows out in a number of different culture medium environments. It should take up to three days to fully understand what we are dealing with."

"OK. I am sure that there will be great interest in what you find, both here and in Mission Control."

38 Memorial

During the next three days that Katya worked on the rock samples, the other crewmembers also were busy. Juliette did a complete analytic check of CARS to ensure that Jango had not made any additional hidden program changes that might endanger the mission. Due to his knowledge of the ERV, Tolya took over Mike's duties of checking the status of the propulsion system and other operational elements on the ship. He reviewed the life support system to make sure that it was not damaged during the depressurization and reboot procedures. Both Tolya and John checked their trajectory and found that the depressurization had nudged the ship a bit off-course, but they were able to correct it using a minimum of fuel.

John communicated with Mission Control about how to deal with the two deaths. They referred the issue to the World Space Council for subsequent action. The decision was made to first notify the friends and families of Mike and Jango, then to report the deaths of the two crewmembers to the press. The news release would say that an accidental depressurization had occurred while Mike and Jango were in the lower deck routinely checking out the spacesuits and the airlock system. It was considered to be a tragic event that would be fully investigated.

However, the possibility of a new life form being discovered on Mars was not communicated to the public, in part because this had not yet been established, and in part because the World Space Council was discussing how to report this kind of unprecedented news. Complicating matters was the fact that there was a homicidal event committed by a crewmember who had suffered a manic episode in space. Being psychotic was one thing, but killing someone as a result of being infected by an alien entity was quite another.

There was widespread mourning on Earth over the two deaths. The activities of the crew of *MarsExplore* had been front-page news for over two years, and the individual crewmembers were well known to billions of people. It was decided that a funeral service was needed to achieve closure. Because the bodies were not available, this took the form of a memorial service. Following naval tradition, John, as the expedition's leader, would conduct it. Care was given to making it non-denominational so as not to offend any particular religious group. John spent the better part of a day scripting the service with people in Mission Control, as well as receiving advice from religious leaders on Earth.

The actual service took place on Sunday, July 15, three days after the deaths. It was beautifully orchestrated and focused on the contributions of Mike and Jango to the expedition, the tragedy of their deaths, and the notion that they were now part of the "infinite spirit of the Cosmos" and would never be forgotten. Since the service was taped and shown worldwide on television, it was estimated that some 4 billion people watched it. It was generally well received, although some fundamentalist groups were annoyed that the word "God" had not been mentioned, and some atheists protested the word "spirit." Nevertheless, the families were satisfied, and closure was achieved over the passing of the two heroes.

When the service was completed, the crewmembers were asked a series of questions about their feelings and reactions to the deaths of their two colleagues via tape-delayed interviews from members of the worldwide media pool. The questions generally related to their feelings about losing valuable colleagues with whom they had worked intensely for several years. The notion that space travel was still a dangerous venture came up, especially related to a first-of-its-kind mission, such as the expedition to Mars. Katya commented on the stressors that the crew experienced on such a mission, such as a tremendous sense of isolation from home and the difficulties in speaking with family and friends in real time as a result of the communication time delay.

John commented on how it felt to be so autonomous from Mission Control, and how well their training prepared them to deal with unexpected events and fend for themselves. Using the analogy of previous sailing expeditions to the New World, he reiterated that danger was part of any new venture into the unknown.

After their final comments were recorded and sent back to Earth, they turned off the transmission equipment. It had been a busy and emotional three days, and they decided to take the rest of the day off. John went up to his sleep pod to take a nap. Katya went back to the lab to check on the rock samples. Juliette and Tolya floated over to the dining area for some tea.

"I feel really sad about the deaths," Juliette confessed. "Granted, Jango may have turned into some kind of alien, and Mike killed him as well as himself, either to save us or because of a delusion. But for nearly four years, they were our friends and colleagues during training and the mission itself. I really miss them."

Tolya took her hand. "Yes, I know how you feel. I miss them too, no matter what rock samples will reveal. They were good crewmates for most of mission. You know, it is ironic. During training, we worried about accident occurring on Mars surface, maybe crashing on landing, or Rover breakdown, or medical emergency. After we took off on trip home, I thought we were safe. I never expected that return would be so eventful."

"Yes, I know. I never expected it either. By the time we get back, things will have normalized, and we will have many tales to tell. But Mike and Jango never will. Both of them worked hard to overcome problems. Mike was an orphan and had to prove himself. In a sense, so did Jango, coming from a farm in rural China, but then ending up going to Mars! No one ever said that space flight would be easy or safe. I know that people back home expect things to go perfectly. Sometimes I do too. But when I am alone, I realize just how dangerous traveling in space can be. Equipment can malfunction, a meteoroid can collide with you from out of nowhere—the unpredictable can happen."

She paused, took a breath, and continued.

"But the unpredictable can be a good thing, too." She took his hand. "You know, finding you is a good thing. I have been working so hard all my life to fly into space that I sometimes neglected other needs that I have. I used to think that being in space was my destiny, my reason to live. Now, on this expedition, I have fulfilled my destiny, but I realize that I have more to live for as well. And that is you."

Tolya was silent, then said: "Yes, I feel same way. I have been action guy all my life, too busy to settle down. But it is time. This will probably be last trip for you and me. But it won't be last time we are together. We can continue to explore new things, but on Earth, with each other."

"I am ready for that now," she responded.

He gave her a kiss, and then they hugged for several heartbeats.

Katya looked up from her microscope and gazed across the mid deck at the two lovers. She smiled and continued her work.

39 Lab Findings

The green rocks certainly had a story to tell. Reconfiguring the BSL-4 cabinet in the lab to allow it to mimic a Martian-like environment, Katya carefully unsealed the rocks and broke off samples for analyses. She especially was interested in the dark, rough smudgy areas that reminded her of lichen growing on terrestrial rocks. She prepared some of the samples as slides and examined them under the microscope. She saw tiny rod-shaped bodies that looked like microorganisms. They resembled the drawing in Jango's diary. She placed other samples into culture tubes that had growth media of different compositions and were incubated under varying conditions of air, temperature, and pressure. She examined these daily. After dinner on the day following the memorial service, she was ready to give her fellow crewmembers a briefing.

"I can confirm that we have found life on Mars, not just fossils but microorganisms capable of living and reproducing. Under the microscope, they look like single-cell organisms. On staining, they do not appear to have a nucleus or any complex organelles, so they are probably primitive forms of bacteria or some new type of microorganism."

After a pause produced no questions, she continued.

"The samples grew best in the cultures where the conditions were most like those found in the Martian hot spring. They especially thrived in hot temperatures when I added water, Mars soil supplement, and high levels of pumped-in carbon dioxide and hydrogen. And guess what gas they produced?"

"Methane," responded Juliette.

"Correct. Just like the formulae Jango wrote in his diary. The microorganisms are methanogens, at least in the wild. I don't know what they might produce in our brain."

John interrupted: "You know, Katya, Mike seemed to be fearful that these microorganisms were somehow turning Jango into something non-human. Could that be possible?"

Katya thought for a moment, then said: "That is an interesting question. Most pathogens on Earth affect our bodies in a way that is seen like a foreign invasion: either we conquer the invaders and get better, or they win and we become ill and die. But sometimes a symbiosis occurs, where both sides benefit or even change as a result of the interaction."

"Kind of like humans and dogs," volunteered Tolya.

Everyone laughed.

"Sort of," continued Katya. "Maybe something like this happened to Jango, where a new kind of synthesis occurred. Perhaps they changed his personality to suit their needs. But also, it is possible that Jango in turn gained new abilities and was becoming something that was more than human, like Mike reported."

She paused for a moment, then continued: "But I am just speculating here, since I didn't have a chance to study the process in action or to speak in detail with Mike and Jango about these issues. They are now dead, and I don't have the equipment in the lab to study the morphology or genetics of our Martian microorganisms in any great detail."

Juliette shuddered, then asked: "Katya, are we likely to be infected during our trip back?"

"I do not think so. The cavern samples are well sealed, and I will store them with our other rock samples."

"OK, Katya," said John, "Prepare a report on this, something that we can send to Mission Control?"

"Certainly, John. I will report the facts as I know them presently. I will leave it to others back home to speculate about what this all might mean for our future."

"I expect that the World Space Council will develop a strategy of dealing with this issue. There will no doubt be a great deal of discussion over these samples for years to come."

40 E-mails

To: john.c.wood@MarsExplore.gov (PERSONAL AND CONFIDENTIAL)
From: asaduddin.mogawi@worldspacecouncil.org
Subject: Crewmember Deaths
Date: Thursday, 19 July 2035, 1000 hours EDT

Commander Wood,

The World Space Council *MarsExplore* Expedition Committee has completed its deliberations on the matter of the deaths of two of your crewmembers: Chief Engineer Michael Lipinski and Expedition Geologist Wang Jianguo. They have also examined your report of possible life forms from Mars that may have infected the two deceased crewmembers.

The Committee accepts your evidence that Chief Engineer Lipinski had a manic breakdown during your return that was not predictable prior to launch. It also accepts the fact that microorganisms were found on Mars that are of alien origin. However, the Committee has concluded that there is no definite proof that either of the deceased crewmembers was infected by these microorganisms. Perhaps both of them were mentally ill with delusions that alien life forms had entered their bodies. Since they both perished in the unfortunate accident involving a rapid depressurization from an open airlock, possibly related to a CARS program malfunction or human error, it is impossible to determine for sure that they were infected. To suggest this without proof would besmirch their good names and create an unwarranted panic scenario of a human take-over by Martian organisms.

Consequently, on the advice of the Committee, the World Space Council is directing you to refrain from issuing any further statements about infectious organisms from Mars. This pertains to communications you may have with family and friends on Earth during your return, as well as to any discussions you may have with the public after landing. You will be debriefed by your assigned flight surgeons and by World Space Council staff during your quarantine period in Houston, which will be lengthened until the issue of your possible exposure to the microorganisms has been resolved. In your debriefs, you will be expected to honestly and candidly report your findings and impressions. But again, this candor should be limited to the assigned space agency personnel and not to relatives, friends, acquaintances, or the media.

Sincerely,
Asaduddin Mogawi
Mission Liaison
MarsExplore Expedition Program

Asaduddin Mogawi
Mission Liaison,
MarsExplore Expedition Program
World Space Council—USA Division
United Nations Headquarters
First Avenue at 46th Street
New York, NY 10017

****

To: john.c.wood@MarsExplore.gov (RESTRICTED USE)
From: swood@hotmail.com
Subject: Hello
Date: Thursday, 19 July 2035, 2314 hours CDT

Dear John,

I hope that you and the rest of the crew have had a chance to unwind a bit after all the excitement. All my friends thought the memorial service was beautiful and that the comments you and Katya made afterwards were quite moving. It sounds like you have all been under a lot of pressure. But I need to tell you something, and I hope that this will not add significantly to your stress. This is very difficult for me. I was going to wait until you returned, but it has been gnawing at me and I haven't been sleeping very well.

I have told you about my art class and the classmate who was very supportive to me. Well, a few weeks ago we went out for coffee to talk, then back to his apartment. I told him how lonely I felt sometime, and the next thing I knew, he was holding me and kissed me. Part of me felt repulsed, but another part of me felt loved and comforted. It didn't go any further than that kiss, and I quickly left. I haven't seen him since. But the event made me realize how alone I feel at times, especially with the knowledge that Melanie will soon be leaving.

John, I don't think I can put up with our life together any more. It seems like you are always gone, missing important events, and I believe the kids have suffered for it. Although they are proud of you, they have at times complained to me about growing up without a father being around. Mel has broken out in tears because you were not here for important events, such as her prom and when her team won the regional soccer tournament. Robert has been more stoic about it, but I know that he has felt the loss as well. And so have I. I feel like I really don't have a life partner.

I realize that due to the radiation build-up, the chances of your flying again are minimal. But you are a celebrity now, and I know that you will be making frequent trips and attending special occasions to discuss your experiences. Our whole family will be in the public eye, and reporters and politicians will be all around us. I want a normal life and not have to share a husband with strangers. Hasn't our family suffered enough for the space program?

Some days I feel that I am better off separated from you. At least my life is stable and predictable, without constantly having to adapt to your comings and goings and having to put on a brave face. I know this might sound cold, but this is the way I feel, and I sometimes hate myself for it.

I think we need to sit down and discuss our future after you return. I am sorry to raise this issue with you now, especially since you are still several months from touchdown. But I think you should know how I feel. And I want to give you some time to think about us and our future during your return. Maybe we will need to see a counselor. Something needs to change.

Steph

41 Problem

John had been obsessing about his last two e-mails. He was angry at both the mission liaison and his wife. But his wife's e-mail also made him feel guilty and scared. And confused. What would await him when he returned? Would his family life be in shambles just at the time that he would be spending the rest of his career on Earth? How could Stephanie do this to him? How could he cause her to be so lonely?

Looking out of his open sleep pod door, he saw Katya emerge from the upper hatchway opening. "Katya, could you come in for a moment"

She floated over to him. "Sure, John."

She entered his pod, and at his bidding she closed the door behind her before strapping herself down on the bed.

"Katya, I have received a troubling e-mail from the World Space Council. They're choosing to discount our report of a Martian infection of Mike and Jango and blame the deaths exclusively on an airlock malfunction, possibly related to an errant CARS program or to some sort of human error. How sure are you about the infection?"

"I cannot prove anything since I do not have blood samples from either of them that show any sort of infection. At last check, their blood counts were normal, their liver functions were normal, and their last physical exams were normal. The hot spring rock samples showed many microorganisms, as I said in my report, but I cannot link these to a definite human infection."

"Well, we're going to take a lot of heat when we get back. If the accident story holds, then all of us will be held accountable. Me as Commander, Juliette as the CARS expert, you as the physician treating a crewmember who was mentally ill and may have caused the accident—they will find ways to implicate all of us."

"I know, but I really have nothing else to contribute."

"Can you offer a professional opinion?"

"It is all in my report."

I wonder what's going on with her? John thought. *She seems distant.*

"Katya, are you alright?"

"Yes, Commander, I am just tired. It has been a difficult time of late."

"I know. For all of us."

"I saw on the log that you received an e-mail from Stephanie. How is she?"

After a pause, he blurted out: "I think she's getting fed up with me and my time away from home. I can't blame her. She told me about feeling lonely and alone, and needing to be comforted by other people. She said she saw another man, not sexually, but for comfort. Being an astronaut is very stressful on family life. I'm not sure what to expect when I get home. I think she wants us to see a couple's counselor."

"I am sorry about your difficulties."

"And she says that my separations have affected the kids as well, implying that I have not been a good father. Furthermore, she feels that they…"

He was interrupted by her yawn.

"Well, anyway, I don't mean to burden you with my personal problems."

"Yes, John. I need to go to my sleep pod to look over the latest crew exercise results. Maybe we can discuss this another time."

She unstrapped herself and floated out the door.

42 Epilogue

The remainder of the trip back was uneventful. The ERV reached the Earth's vicinity on time, and the orbital insertion was successful. The landing occurred as planned on November 22, 2035, Thanksgiving Day in the United States. The President gave a speech on worldwide television, and billions of people rejoiced at the accomplishment.

The crewmembers went into an immediate quarantine until it could be assured that they were not carrying anything that might contaminate the Earth. The Martian soil and rock samples were sent under strict isolation to the large BSL-4 facility in Galveston, which had been specially adapted to deal with these samples. It was expected that the analyses of the samples would take months or even years to complete, especially since they were anticipated to contain live alien microorganisms.

The World Space Council had decided to run the story of "possible microbial life from Mars." However, this story did not mention crewmember infections, nor did it describe any possible role of the microorganisms in the deaths of Mike and Jango. Their demise was attributed to an unfortunate accident involving the airlock. Although CARS was implicated, human error by the two involved individuals could not be ruled out. The surviving crewmembers were not reprimanded, however, since the errors that were committed did not seem to be a result of their negligence. The Safety Division of the World Space Council formed a subcommittee to examine the possible mechanical or programmatic errors that could have contributed to the accident, and they were tasked with developing a fail-safe procedure so that a similar accident would not occur in the future. The *MarsExplore* crew signed off on this storyline, as did the involved space agencies.

After being released from quarantine, the four crewmembers went on a worldwide tour. They were heroes! Everyone knew their names. Although initially caught up with all of the excitement, the four of them soon began to struggle with the burden of the fame and glory they were receiving. Everywhere they went, people wanted a piece of them: politicians soliciting votes for reelection, talk show hosts looking to improve ratings, magazines and newspapers wanting to sell more issues. There was talk of product endorsements: tee shirts, hats, cereals, soft drinks—all named for the crewmembers or Martian topographical features. But since the four of them worked for space agencies and were under the control of the World Space Council, these approaches were resisted as tainting the image of heroes who were considered to be world icons. But perhaps later, after all the uproar quieted down, and they retired from the government agencies, money could be made to supplement their pensions.

∗∗∗∗∗∗∗∗∗∗∗∗

John especially was affected by all of the hoopla. Six months after landing, he remained a troubled man. Two crewmembers had died under his command, and most people did not know what really happened. There was a lingering undercurrent of blame. How did the airlock blow open accidentally? Weren't there safeguards to prevent such an event? As Commander, shouldn't he have made sure that people followed the appropriate protocols?

Furthermore, after several years preparing for and participating on this expedition, his home life had suffered. Stephanie had developed new interests and had learned to manage things in his absence, although she resented the lack of support she felt from John. Both of his children had grown up, and he had missed the joys of seeing them mature into young adults. Beyond this, there was the constant recognition, the calls from the news media, and the expectations from NASA that he would participate in all of their Mars public relations tours. He realized that he could never again fly in space, and he rued the thought that his flying career was over. He started drinking too much. He and Stephanie fought too much. After a trial separation, they entered couples therapy, which helped bring them back together. But the readjustment was difficult, more so than he had expected. He needed to redefine his life and gain new meaning regarding his future. In many ways, his re-entry back into his old world was more difficult than his entry into the new world that was Mars. He missed his time on the Red Planet—things were simpler then!

****

Juliette and Tolya also had adjustments to make. Six months after their return, when things had quieted down a bit, they were offered a rent-free apartment together in Moscow. This resulted in a media feeding frenzy. They weren't a normal couple: they were astronauts living together who were not married! In addition, it created a strain for Juliette, since she still had responsibilities with ESA and had to make frequent trips back to Paris. So, despite being committed to each other, they in fact were apart more than they would have wished. At times, she would book a flight back to Russia to be with Tolya, then have to leave the next day to return to France to meet yet another obligation. It made for a chaotic relationship but resulted in brief but passionate rendezvous.

And the media loved it! They became the darlings of the press. The handsome Russian pilot and the beautiful French computer expert were trailed by paparazzi everywhere. Gossip columnists had a field day. Would they get married? Were they having children? There was even talk of a movie being made about the expedition that would focus on their romance in space and downplay the historical and scientific achievements that were made. However, they managed to stay together through all of this attention and continued to define and relish their relationship.

****

Katya likewise received a great deal of attention after she returned from the expedition, and she took steps to establish a more stable and quiet life. Two months after her quarantine ended, she resigned from the Russian Space Agency and accepted a microbiologist research position at her old university in St. Petersburg. She moved into a small apartment just a block from Nevsky Prospect, which was bitter sweet for her since this was the street on which Valya had been killed. Although she was asked to give interviews and presentations regarding her experiences on Mars, she politely declined the offers. She had a new mission, and it demanded a low profile, which she was able to achieve as an academic working in a laboratory.

The change came on slowly. During the latter part of her return home, she often felt tired and a bit confused. Her mind wandered at times, and occasionally her thoughts seemed foreign. After returning to Earth, she had a vague sense of being different. She thought that this was simply due to re-adjusting from microgravity, but then she realized that more was going on. It must be related to the green dust she inhaled in Jango's sleep pod. She had a growing belief that she was unique and had a special purpose in life. Her research would take on a new focus based on this belief. The secret was still in her blood. It was undetected by the standard laboratory tests that were done as part of the routine re-entry physical examinations. Special preparations would have been needed to isolate it: Mars-like soil, a little water, hydrogen gas, and carbon dioxide. Methane was the marker of discovery.

She could still think like a human as she integrated the new ideas into her personality. She began to marvel at the future, as her thoughts became more and more clear:

My progeny will be passed on. They are my children. They will grow and prosper in their new hosts. The result will be human-like but not human, something different, yet familiar. Destiny has been served. As in eons passed, they have come to a new home. But in reality, they are coming back home. They are the new Martians, but their home is Earth.

Part II

The Science Behind the Fiction

This constellation map is from the 1782 edition of Johann Bode's star atlas *Vorstellung der Gestirne*. It shows Perseus, carrying the monstrous head of Medusa, who is about to rescue the chained Andromeda from Cetus, the sea monster (who is located elsewhere in the heavens). Whether they are battling aliens or the psychosocial stressors inherent in a long-duration, isolated expedition to Mars, male and female crewmembers will have to get along and look out for one another. Courtesy of the Nick and Carolynn Kanas Collection; and *Star Maps: History, Artistry, and Cartography, 2nd Edition.*, Nick Kanas, Springer/Praxis, 2012.

Psychosocial Issues during an Expedition to Mars

In *The New Martians,* the crewmembers undergo a great deal of psychological and interpersonal stress during their return home, in part prompted by the actions of a mysterious presence on board. Of course, no one knows for sure if such a presence will actually materialize during a real Mars expedition! But psychosocial issues will nevertheless affect a Mars crew due to the isolation, confinement, and long separation from family and friends that will characterize such a mission. In what follows, many of these issues will be reviewed, followed in each section by illustrations from the novel.

1 Psychological and Interpersonal Stressors On-orbit

Much is known about the psychosocial issues that affect astronauts in space from anecdotal reports, scientific studies performed in space analog environments on Earth (e.g., Antarctic, space simulation chambers, submersibles), and experiments conducted on missions involving orbiting spacecraft. Although some of this information can be extrapolated to a Mars mission, one must be cautious in taking this leap from the Earth and its environs to a planetary body deep in space, since new stressors will affect the crewmembers that are not present closer to home. But we must begin somewhere, so let's first examine some of the psychological and interpersonal stressors in the near-Earth environment [1]. These are summarized in the left hand column of Table 1.

Isolation and confinement occur in all space missions and force the crewmembers to interact together in a small space far away from home. Danger is part of the mission, whether from a micrometeoroid impact, a malfunction in an important piece of equipment, a fire, or any number of other factors. Monotony can also occur, although there are also periods of high workload during spacewalks and emergencies. Personality conflicts can interfere with

Table 1   On-orbit and planetary psychological and interpersonal stressors

On-orbit stressors	Additional stressors on a Mars mission
Isolation and confinement	Cultural issues
Possible danger	Selection Issues: who will go?
Monotony	Effects of long-term microgravity
High workload (e.g., spacewalks, emergencies)	Effects of long-term radiation
Personality conflicts	Extreme isolation and loneliness
Crew size	Dependence on machines and local resources
Time effects	Limited social contacts and novelty
Leadership roles	Leisure time
Crew-ground communication	Lack of support due to communication delays
Crew heterogeneity	Increased autonomy
Common language	Earth-out-of-view phenomenon
Cultural issues	Family problems at home

crew cohesion and performance. The best crewmembers for long-duration space missions are those who are comfortable working alone on an activity when diligence is called for, yet at the same time are team players who enjoy relating with their colleagues during meal and leisure times [1]. Crewmembers undergo up to two years of pre-launch training, although currently there are relatively few opportunities for behavioral health experts to observe them for signs of incompatibility before they are sent into space. The number of people comprising a crew may be important. In studies of unstructured groups on Earth, it has been shown that odd-numbered groups form a consensus better than even-numbered groups, and larger groups are more cohesive than smaller groups, since people can usually find a person or two with similar interests to counter feelings of isolation [2].

In missions lasting six weeks or more, time effects have been observed. For example, crewmembers in space or in space analog environments have been found to exhibit significant psychological and interpersonal difficulties after the halfway point of their mission. The idea is that some crewmembers arrive at this milestone with relief that things are going well, only to realize that there is still another half to go before they will be home. It is not so much the number of days that have transpired, but the perception of "halfway" that has the most psychological relevance. Some investigators have described the presence of a "third quarter phenomenon," characterized by increased homesickness, depression, irritability, and decrements in crew cohesion shortly after the half-way point [3].

To examine time effects, our research team conducted two NASA-funded international studies of psychological and interpersonal issues during a series of on-orbit missions lasting 4–7 months to the Mir and the International Space Stations. The Mir study sample consisted of 5 American astronauts, 8 Russian cosmonauts, and 42 American and 16 Russian Mission Control personnel. The ISS study sample consisted of 8 American astronauts, 9 Russian cosmonauts, and 108 American and 20 Russian Mission Control personnel. Subjects completed a weekly questionnaire that included items from a number of well-known measures that assessed mood and group dynamics. Both studies had similar findings, so in a sense they replicated each other. In both studies, there were no significant changes in levels of emotion and group interpersonal climate over time. Specifically, there was no evidence for a general worsening of mood and cohesion after the halfway point of the missions, and no evidence for a third quarter phenomenon [4, 5]. It should be noted that some individual crewmembers showed evidence for such a decrement just after the halfway point of their missions, but others showed no such effect or even experienced an improvement in emotional state during the second half. Our belief is that the absence of general negative time effects in our studies was the result of supportive actions taken by flight surgeons and psychologists in Mission Control, which included the sending of favorite food and surprise presents on resupply ships and increased communication with family and friends on Earth. Such actions helped to provide novelty and counter the effects of isolation, loneliness, and limited social contact. The celebration of mission milestones and holidays likely contributed to the maintenance of morale as well.

Research in the Antarctic and other isolated and confined environments on Earth suggests that the identified leader has at least two major roles in a group [1]. The first deals with setting the agenda and getting the work done: the task role. The second deals with supporting the team and paying attention to group morale: the support role. These aspects of leadership become especially important at different times during the mission. For example, during emergencies, the task role is crucial, whereas during monotonous periods, the support role becomes more relevant. Ideally, the commander of a space mission is comfortable with both roles and knows how and when to use them. We studied the impact of leadership roles on group cohesion in both our Mir and ISS studies. We found that the support role of the mission commander was significantly and positively related to group cohesion among crewmembers. In our Mission Control subjects, both the task and support roles of the team leader were significantly related to cohesion in the ground-based work groups [4, 5].

During on-orbit and lunar missions, the communication between the crewmembers and people on Earth is very important for morale. In a survey of 54 astronauts and cosmonauts who had flown in space, Kelly and Kanas found their respondents to rate a sense of shared experience and a mutual excitement for space flight as two factors that significantly helped their communication with Mission Control personnel [6]. They further acknowledged the value of contact with loved ones on the ground as having a positive influence on mission performance. Crew heterogeneity is also an important factor. Space missions typically involve people of both sexes, diverse professional and experiential backgrounds, and different life experiences. In the long run, such diversity can be beneficial, since it provides novelty and stimulation later in the mission when people begin to tire of the routine and look for something new to talk about. However, diversity can also be stressful, especially initially when people are adjusting to individual differences. One important counter to the negative effects of heterogeneity is for all of the crewmembers to fluently speak a common official mission language. This not only enhances efficient communication of ideas during work activities and emergencies, but it also improves bonding with fellow crewmembers through a better understanding of the connotations of their speech and the meaning behind their comments and jokes. In their survey, Kelly and Kanas found that 100 % of the respondents acknowledged that it was important for space crewmembers to be fluent in a common language, with 63 % rating it as "very important" [7].

Stuster examined a number of interpersonal stressors that affected space crews working on-orbit [8]. He performed a content analysis of personal journals from ten ISS astronauts that were oriented around issues that had behavioral implications. He found that 88 % of the entries dealt with the following categories: Work, Outside Communications, Adjustment, Group Interaction, Recreation/Leisure, Equipment, Events, Organization/Management, Sleep, and Food. The crewmembers reported that their life in space was not as difficult as they expected prior to launch, despite a 20 % increase in interpersonal problems during the second half of the missions. It was recommended that crewmembers be allowed to control their individual schedules as much as possible.

Examples from the Novel: In *The New Martians*, the crewmembers have been away from Earth for two years. All the excitement from their exploration of the Martian surface is past, and they are tolerating the boredom of the long return home. By now, they have adjusted to each other's personality quirks. Fortunately, nearly everyone has at least one other person on board who shares some demographic or work-related characteristic to give them support. It has helped for the crewmembers to be fluent in English (the designated mission language), although Juliette, the French computer engineer, struggles with some words, and Tolya, the Russian pilot, speaks with an accent. To avoid monotony, the crewmembers look for things to do to occupy their time. Tolya especially has few critical tasks to perform since his piloting skills won't be used until they get closer to Earth, and he experiences some boredom. John, the commander, is sensitive to the ennui that has enveloped the crew and is exercising his supportive leadership role. Together with Katya, the mission physician, they monitor the crew and encourage stimulating activities, such as celebrations and parties. John also finds it useful to write e-mails home to his family. In one discussion, Katya reminds John that some people working under isolated conditions may experience depression after the halfway point of their mission. This seems to be the case for this crew, until a number of dangerous events occur that shake them out of their monotony and produce periods of high workload that are needed to overcome the emergencies.

2 Cultural Issues

A particularly important issue during all multinational space missions per-tains to cultural issues. People from different national groups interact differ-ently and have certain expectations from other people. Take, for example, cul-tural norms. People from Mediterranean countries are typically behaviorally animated and comfortable being physically close to one another when speak-ing, whereas people from Northern European countries are more reserved and have less tolerance for someone gesticulating close to them, which they may perceive as boorish behavior or as a sign of aggression. During space missions, organizational culture also is important. Space programs vary in their degree of formality and their dependence on procedures and redundant equipment located on board (e.g., American space program) versus simply calling in ex-perts on the ground to resolve problems and suggest on-the-spot repairs (e.g., Russian space program). An astronaut or cosmonaut used to one system may have difficulty adapting to another.

We found some cultural differences in our Mir and ISS studies. Crewmem-bers scored higher in cultural sophistication than Mission Control person-nel. Russians reported greater language flexibility than Americans. Americans scored higher on a measure of work pressure than Russians, but Russians reported higher levels of tension on the ISS than Americans [9, 10].

Other research teams have also taken a look at how cultural factors affect space travel. Tomi et al. examined potentially disruptive cultural issues affecting space missions in a survey of 75 astronauts and cosmonauts and

106 Mission Control personnel [11]. The subjects rated coordination difficulties between space organizations as the biggest problem. Other problems included communication misunderstandings and differences in work management styles. Sandal and Manzey surveyed 576 employees of the European Space Agency and found a link between cultural diversity and the ability of people to interact with one another [12]. Especially important were factors related to leadership and decision-making.

Examples from the novel: The *MarsExplore* crew consists of two Americans (John and Mike), two Russians (Tolya and Katya), one French woman who spent time working in the United States (Juliette), and a Chinese man (Jango). Although this crew makeup was determined by a number of political, budgetary, and work-related factors, there is a reasonable balance in terms of cultural diversity, with nearly every crewmember having at least one person who could identify with his or her cultural background. A notable exception is Jango. Not only is he the only Asian, but his selection was strongly influenced by the desire to placate China so that this country would participate rather than compete with the mission. Adding the facts that Jango is a social introvert and had a very modest upbringing, he is clearly in a position to feel culturally isolated and even be scapegoated by the other crewmembers. It would have been helpful if another Asian crewmember, perhaps from China or Japan, could have been included in the expedition, but the mission tasks and the engineering demands of the vehicles restricted the crew size to six.

3 Psychological and Interpersonal Stressors Unique to a Mars Expedition

Specific stressors related to a long-duration planetary expedition, such as to Mars, are listed in the right hand column of Table 1. Again, cultural issues continue to be very important, since such a mission likely will be multinational. Nechaev et al. surveyed 11 cosmonauts regarding their opinions of possible psychological and interpersonal problems that might occur during a Mars expedition [13]. They found the following factors to be rated highly: isolation and monotony, distance-related communication delays with the Earth, leadership issues, differences in space agency management styles, and cultural misunderstandings within the international crew.

In terms of crew selection, not everyone in the astronaut corps will volunteer to be away from family and friends for a two to three year mission, so this may skew the selection process to specific types of individuals (e.g., single people or people without small children). Little is known about the physical, cognitive, and psychological impact of long-duration microgravity and high radiation in deep space, as well as the 38 % Earth gravity that the crew will experience on Mars. The crewmembers will be millions of miles away from Earth, and this will increase their sense of isolation and loneliness to levels higher than in any previous space mission.

People on a Mars expedition will be heavily dependent on their computers and other machines on board for basic life support and operational activities, such as navigation and propulsion. The psychology of this dependence and the ergonomic characteristics of the human-machine interface are important issues to be considered in designing the space vehicles that will be used. Since not all supplies and fuel can be stored on board, the crew will need to depend on local resources in the atmosphere and the surface of Mars to chemically generate water and fuel for the return home. So again, the ease of use and reliability of the relevant equipment will be critical.

Direct human contact will be limited to just the crewmembers, and ennui may result from the lack of novelty and the predictability of interacting with the same people for years. This will make leisure time activities important, and provisions will need to be available to cover a variety of free time activities. These will also need to be flexible enough to account for changing interests.

With the long distances involved in a Mars expedition, delays in crew-ground communication and the inability to send needed resupplies in a timely manner will seriously impair the kind of supportive morale-enhancing activities that occur during on-orbit missions. Given the separation in time and space from Mission Control, there will be high crew autonomy, and the crewmembers will need to be trained to develop their own work schedules and deal with operational and medical emergencies themselves.

Our research team studied the effects of high autonomy and communication delays in three space analog environments on Earth: a submersible vessel located off the coast of Florida, a remote location in Canada where a group of people simulated a Mars exploration, and a Mars analog habitat in Russia [14, 15]. Based on these three studies, we concluded that high work autonomy (where the crewmembers planned their own schedules) was well-received by the crews, mission goals were accomplished, and there were no adverse effects. During the high autonomy periods, crewmember mood and self-direction were reported as being improved, but in one setting Mission Control personnel reported more anxiety and work role confusion. Another research team (Roma et al.) studied the effects of autonomy in groups performing space-related interactive tasks on computers, and they also found evidence that suggested these groups functioned well under conditions of high autonomy [16].

No human being has ever been in the position of viewing the Earth as an insignificant dot in the heavens, the so-called "Earth-out-of-view phenomenon" [1]. The impact of seeing your family and friends and the location of your birth and upbringing reduced to that distant dot may enhance the sense of isolation and homesickness. It is also possible that more profound effects will occur, such as depression, psychosis, or even suicidal thinking. We must be prepared for the occurrence of such reactions to this unprecedented event.

One important issue related to family life is how to inform an astronaut of bad news from home. In their survey, Kelly and Kanas reported that 18 respondents were of the opinion that negative information (such as a death in the family) should be withheld until a space traveler completed the mission, whereas another 22 stated that it should not be withheld [6]. Five additional respondents gave no clear opinion but volunteered that information could be withheld on short-duration space flights but perhaps should be disclosed during long-term missions. Kelly and Kanas opined that when disclosed, bad news from home should be tempered with support and should probably be delayed until after the completion of a critical mission activity.

Examples from the novel: The *MarsExplore* crewmembers were highly motivated to participate in this mission, although Jango had some reservations and John began to realize the toll his frequent absences were taking on his family. Nevertheless, the crewmembers adapted reasonably well to the mission and to each other. Moving in microgravity was sometimes turned into a social activity (e.g., microgravity dancing during parties), and accumulated radiation exposure was controlled but still resulted in this being the last space mission for the crewmembers, a fact they had adjusted to.

One issue that bothered several of them was giving up so much control of life support and operations to CARS, the central computer. This dependence on a machine affected the psychological reactions of the crewmembers. Sometimes they referred to CARS with a masculine pronoun, as if it was a person. At other times, they suspected CARS of having independent motivation and malevolent intent. This idea was fostered to explain a number of computer problems, but still the human-machine interface was an issue throughout the mission. Machines were also helpful in filling leisure time, such as with Jango and his computer games and John and his baseball rankings. But other leisure time activities did not depend on machines, such as Juliette's knitting and the chess game and its aftermath involving Tolya and Juliette! Machines were useful in communicating with people back on Earth, despite the delays it took for verbal or e-mail messages to traverse the long distances involved.

This communication delay was very frustrating, especially for John as he tried to maintain his ties with his family and when he needed advice from Mission Control. The crew had a great deal of autonomy, but this did not seem to be an issue for them. The Earth-out-of-view phenomenon was dealt with by having a telescope on board with which the crewmembers could clearly see Earth and its features. This was a popular activity, especially for

Mike. Two major family problems occurred: the death of Katya's sister, and the "Dear John" e-mail the commander received from his wife. Both events were communicated to the affected crewmembers during non-critical times, and they turned to fellow crewmates for emotional support.

4 Psychological Impact of a Mars Expedition

Astronauts react to psychological and interpersonal stressors in a variety of ways, which are summarized in Table 2. In terms of psychological and psychiatric reactions, the most common relate to simply adjusting to being in space. Living and working under the conditions of microgravity, isolation, and separation from family and friends at home is a novel experience, and most people need time to accommodate. Typically, this occurs within the first few weeks after arriving in space and becoming oriented to the new environment. But some astronauts have experienced clinical symptoms such as anxiety or depression as part of their adjustment. For example, one astronaut beginning a long-duration space mission reported symptoms of clinical depression due to the isolation he felt on-orbit and his separation from his wife [17]. Most of the time, such symptoms go away with support from crewmates and familiarity with the new surroundings. Rarely, a brief course of tranquilizers or counseling from the ground may be needed.

Russian flight surgeons and space psychologists have identified a serious form of adjustment reaction called asthenization, which is related to a psychiatric disorder called neurasthenia [18]. This disorder is defined as a weakness of the nervous system that produces fatigue, irritability and emotional lability, attention and concentration difficulties, restlessness, heightened perceptual sensitivities, palpitations and blood pressure instability, physical weakness, and sleep and appetite problems. Although first described as a neurotic condition in the late 1800s by the American George Beard, there is controversy as to the existence of neurasthenia, since it is not recognized in the current American psychiatric diagnostic system. Instead, its symptoms are incorporated in diagnoses related to anxiety and depression, such as adjustment disorder or chronic fatigue syndrome. However, neurasthenia appears in the diagnostic system used in Europe, China, and Russia. Russian psychologists and flight surgeons have viewed asthenization as an adaptation that affects most cosmonauts going into space, and they have developed a number of countermeasures to deal with it.

Our research team looked for the presence of asthenization by conducting a retrospective analysis of our Mir and ISS data to see if there were any correlations between the results from our mood measures and scores developed by six

Table 2   Effects of psychological and interpersonal stressors on space crewmembers

Psychological/psychiatric	Interpersonal
Adjustment disorders	Group tension and loss of cohesion
Asthenization	Withdrawal and territorial behavior
Somatoform disorders	Lack of privacy and personal space
Psychoses: schizophrenia, bipolar disorder	Scapegoating/subgrouping
Suicidal/homicidal Intent	Displacement
Post-return problems	Sexual attraction/tension

space experts who had experience treating this syndrome [18]. We found that all of the negative mood scores were significantly lower than the scores from the expert prototypes, suggesting that there was no evidence for the existence of clinical asthenization in our samples. However, in a separate analysis we hypothesized that Russian crewmembers would experience a significant correlation between fatigue (the hallmark of asthenization) and depression, whereas American crewmembers wouldn't show such an association [19]. Instead, they likely would experience depression in the context of anxiety, which supports a culture-bound pattern of mood that is consistent with the American model of neurotic depression. Our results confirmed these associations and suggested that patterns of mood states in space crewmembers may reflect national cultural norms. Further work in this interesting area needs to be done.

Somatoform (psychosomatic) disorders also have been reported from space. These consist of distressing physical symptoms that suggest the presence of a medical condition but which are not fully explained by a real physical problem. Instead, they are due to underlying psychological issues. For example, a cosmonaut wrote in his diary that he experienced tooth pain following anxious dreams he had of a tooth infection and his concern that nothing could be done about such an infection while he was on-orbit [1].

Problems related to psychotic conditions (e.g., schizophrenia, bipolar or manic-depressive disorder), which are thought to have genetic or familial determinants, or to suicidal and homicidal intent have not been reported during space missions. This probably reflects the success of the psychiatric screening that is done on people applying to be astronauts or selected for important space missions. However, such severe psychiatric disorders have been reported in astronaut applicants, as well as in up to 5% of people working in some space analog environments [1].

Post-return personality changes and psychiatric problems have affected returning space travelers. Some astronauts have experienced humanistic or even spiritual changes, coming home as more sensitive, people-oriented individuals as a result of seeing the Earth from space without national boundaries and as the common homeland of all of its inhabitants. But for other astronauts, integrating back to their normal life following the excitement of their mission and their new-found fame can be stressful. Some have reported symptoms of major depression, anxiety, and alcohol abuse that have necessitated psychotherapy and psychoactive medications [20]. Depression also affects the spouses of people flying in space, both during and after the mission, and family reentry of the returning spouse may be difficult. For example, two classic studies of male submariners reported that many of the wives had learned to adjust to the absence of their sailor husband when he was on sea patrol, but over half experienced depression and marital strife after he returned and tried to reinsert himself back into the family dynamics. This has led to the expression "submariners' wives syndrome" [21, 22].

Examples from the novel: The crewmembers needed to make a number of psychological adjustments to both the monotony and the crises that occurred during their trip home. As a Russian-trained flight surgeon, Katya was especially sensitive to symptoms of asthenization that could impact on the crewmembers. Tolya also recalled having had psychosomatic symptoms during a previous mission that were related to unconscious factors. But the major psychological problem affecting this crew was Mike's manic breakdown. Since he was adopted, there was no family history to alert mission planners to this possibility. Also, the onset of his first psychotic experience at age 38 was atypical but not unusual for mania. His breakdown caught the crew by surprise, and they had to deal with his behavior in a make-shift manner, since the facilities and protocol for dealing with a psychotic crewmember had not been a major priority for the mission planners, as Katya was well aware. Finally, the Epilogue of *The New Martians* follows the returning crewmembers post-return and shows their reactions to the stressors of the homecoming, including dealing with newfound fame and glory and trying to reunite with a family that has been severely stressed by an astronaut's absence.

5 Interpersonal Impact of a Mars Expedition

Table 2 also lists a number of interpersonal effects related to the stressors that might occur during an expedition to Mars. Top on the list is group tension and its negative effect on crew cohesion. This can result in crewmember withdrawal and territorial behavior, where arguments and even fights can occur over someone intruding into another person's physical space or borrowing something that belongs to him. This becomes even more severe when provisions are not made for crewmembers to have private spaces to store their belongings and retreat to when the need arises. At the group level, people who are perceived as being different can be scapegoated and blamed for the crew's interpersonal problems. Alternatively, subgroups can form, where people ally with like-minded individuals and blame all their troubles on other subgroups. For example, in some early polar scientific missions, the scientists and operational personnel formed two different subgroups that competed for control of the mission activities [1].

But crew tension can also be directed outwardly. People working for long periods of time under isolated and confined conditions may displace tension from their "in-group" to a convenient "out-group" that is more distant and less able to retaliate. Many of us experience displacement at the individual level when we get angry with our boss but cannot confront him or her directly, then go home and yell at innocent bystanders, such as our spouse or a neighbor. Although offering temporary release of emotions, displacement does little to resolve tension in the long run, and it can produce miscommunication and conflicts with outsiders.

Based on a review of early space analog studies, Bill Feddersen and I predicted in 1971 the possibly that displacement would also be found in space crews [2]. Later, as missions became longer and orbiting space stations began to be built, crew debriefings and diaries suggested that this was indeed the case, with crews experiencing intra-group tension and displacing it outwardly to Mission Control, creating serious crew-ground problems in communication [23, 24]. In one mission, this may even have contributed to the refusal of the crew to perform tasks directed from the ground [25, 26].

The notion of displacement has received some empirical support. Based on earlier pilot work, our research team identified six tension and mood subscales in the measures we used in our Mir and ISS studies, and we predicted that these subscale scores would correlate negatively with scores from a measure of perceived support from outside personnel in Mission Control. We reasoned that if crewmembers were experiencing high group tension, they would displace their mood state onto Mission Control staff and perceive them as not being very supportive. As predicted, all six correlations were significant and in the predicted negative direction [4, 5]. In Russia, Gushin and his colleagues studied cohesion and in-group/out-group conflicts in isolated groups located both on the ground and in space using an analysis of speech patterns. They found that over time, the crewmembers showed decreases in the scope and content of their communications and a filtering in what they said to outside personnel, which they termed "psychological closing" [27, 28]. Sometimes this served to hide medical and psychological problems in the crew. They also tended to withdraw into themselves and became more egocentric, a process the investigators called "autonomization." In a displacement-like manner, these factors resulted in some members of Mission Control being perceived negatively as opponents. This research team also found that crewmembers became more cohesive by spending time together (including joint birthday celebrations) [29], and that the presence of subgroups and outliers (e.g., scapegoats) negatively affected group cohesion [30]. Finally, in a closed self-sustaining facility on Earth that was viewed as a long-duration human space exploration analog (Biosphere 2), eight people were confined for two years [31]. They experienced interpersonal conflicts, ultimately breaking up into two factions, and they perceived an insufficient level of support from outside monitoring personnel.

From June 2010 to November 2011, a unique ground-based space analog mission took place in Moscow that was called the Mars 500 Program [32]. It was designed to simulate a 520-day round trip expedition to Mars. The mission included periods of time where the isolated crew functioned under high autonomy conditions, including communication delays with people working outside in the simulated Mission Control. Six men were confined in a facility that was located at the Institute for Biomedical Problems in Moscow. The lower floor consisted of living and laboratory modules for the international crew, and the upper floor contained a mock-up of the Mars surface on which the crew conducted simulated geological and other planetary activities.

Several psychosocial studies were conducted during the actual 520-day mission. Gushin et al. found changes in crewmember time perception, evidence for the displacement of crew tension to Mission Control, and decreases in crewmember needs and requests during high autonomy, which suggested that they had successfully adapted to this condition [33]. Sandal reported that the crew exhibited increased homogeneity in values and more reluctance to express negative interpersonal feelings over time, which suggested a tendency toward "groupthink" [34]. Van Baarsen et al. found that the crewmembers experienced increased feelings of loneliness and decreased support from colleagues over time, which negatively affected cognitive adaptation [35], and that several factors affected motivation [36]. Basner et al. used wrist actigraphy, the psychomotor vigilance test, and various subjective measures to study the crew and found a number of individual differences in terms of sleep pattern, mood, and conflicts with Mission Control [37]. The investigators also found a tendency for active wakefulness levels to drop throughout the mission until the last 20 days, when they rose in anticipation of the end of the seclusion [38]. Finally, Tafforin evaluated fixed video recordings of crew behavior during breakfasts and found variations in personal actions, visual interactions, and facial expressions, but a general decrease in group collective time from the outbound to the return phase of the simulated mission [39].

One final interpersonal issue deserves mention. In a multiyear space expedition involving sexually active men and women, it would be expected that sexual attraction and tension (both physical and psychological) would exist [40]. Anecdotal reports from space and studies of isolated groups on Earth suggest that some male-female teams experience disruptive sexual jealousies and rivalries, but most groups show evidence of being stable and performing equal to or better than all-male teams [1, 41–43]. A compatible personality mix and diligent pre-launch training seems to positively affect later group cohesion. In terms of sexual activity, there have been no confirmed episodes of heterosexual (or, for that matter, homosexual) encounters in space, according to American and Russian astronauts and space experts [44, 45]. Due to the lowered testosterone levels and upper body fluid shifts that occur in microgravity [41, 43, 46], there has been some question raised about the ability of men to become aroused. However, one astronaut reported that most days that he was in space, he (and at least one other colleague) had an erection upon awakening [47], and a male space tourist reportedly had a similar experience [40]. Another sexual issue relates to being able to remain intimately in contact given the physical effects of microgravity on orientation and movement. Space experts believe that couples could adapt to this situation [44]. Like Tolya and Juliette in the story, this might require creative positioning and restraint [40] or the assistance of a specially-designed "2suit" to keep the couple enclosed and together [48]. If sex does occur in space, it is important that women avoid pregnancy, since there is evidence from animal studies that fetal and neonatal development may be negatively affected by microgravity [41, 42]. Because many female astronauts take oral contraceptives during their mission to regulate menstrual cycle function and attenuate bone loss and osteoporosis, [46] pregnancy for these women would be a low risk.

Examples from the novel: The crewmembers were under a great deal of tension that tested their ability to form a cohesive group. With the exceptions of Jango, who often withdrew to avoid interactions with his fellow crewmates in space, and Mike, whose progressive manic breakdown caused him to become isolated and territorial, the other members of the crew were able to cohesively problem-solve the various crises that occurred. The special relationship between Tolya and Juliette caused them to form a "couple" subgroup, but they still interacted with their crewmates, who if anything were tolerant and happy for the two of them as their relationship developed. Only Mike expressed feelings of jealousy, but this was compounded by his psychological problems. There was evidence for the displacement of tension to personnel in Mission Control, especially by Mike who was very critical of their activities. But all of the crewmembers were mindful of the fact that they had to deal with acute problems and emergencies that occurred on board themselves, and at best

Mission Control could only provide an advisory role, sometimes long after a crisis had passed due to the communication time delay.

6 Positive Effects of Being in Space

Isolated and confined environments can be growth-enhancing. Several writers have discussed the salutogenic, or health-promoting, reactions some people have to the adverse conditions found in polar environments and in space [49–51]. These include increased fortitude, perseverance, independence, self-reliance, ingenuity, comradeship, and decreased tension and depression.

There has been some empirical evidence supporting the positive aspects of space travel. Our research team surveyed 39 astronauts and cosmonauts and found that 100 % of the respondents reported positive changes as a result of flying in space [52]. One factor significantly stood out: Perceptions of Earth. An item in this subscale that dealt with gaining a stronger appreciation of the Earth's beauty had the highest positive mean change score. This should not be a surprise, since gazing at the Earth from space has long been reported to be a favorite activity of astronauts [23, 24].

Extending pioneering research begun in the early 1990s, Suedfeld and his colleagues content analyzed the published memoirs of 125 space travelers [53]. As a result of being in space, astronauts and cosmonauts reported more Universalism (i.e., a greater appreciation for other people and nature), Spirituality, and Power. There were also cultural differences: Russian space travelers scored higher in Achievement and Universalism and lower in Enjoyment than Americans.

Examples from the novel: During the Mars expedition, Tolya and Juliette developed a loving relationship and were prepared to continue it back on Earth. Although the separation was difficult for his family, John learned many things about himself and the changes he would need to make in the future to deal with family problems. Despite his breakdown, Mike gained an appreciation of several things: the beauty of the Earth by observing it through the telescope, the specialness of the mission and its impact on his life, and his attachment to his fellow crewmembers. Jango and Katya also had experiences that they felt were unique and special, although perhaps not typical of most space missions!

7 Life on Mars and Panspermia

The issue of finding life on Mars had a major psychosocial impact on the *Mars-Explore* crewmembers. In fact, the search for life was a major mission goal, and the belief that they had failed was a major disappointment and source of intra-crew conflict. In reality, their disappointment is understandable. We know from orbiting spacecraft and surface probes that Mars currently has frozen water trapped in its polar caps and just below its surface. The presence of gullies and other geological features suggests that liquid water flowed relatively recently on Mars. As mentioned in the story, the *Tharsis* region is thought to be a volcanic area, and probes have imaged openings suggestive of caves, so it is not impossible for sheltered hot springs to be present inside that contain steam and liquid water. Mars' atmosphere, although thin, contains carbon dioxide and methane, so it is possible that under the right conditions microorganisms that give off methane could be present. Methanogens exist naturally on Earth [54], and they and other extremophiles have been grown in the laboratory under partially analogous Mars-like conditions [55–57].

If found, could such microorganisms be infectious? Katya speculated that the symptoms of the disorder that affected Jango and Mike reminded her of Creutzfeldt-Jakob disease in humans and Mad Cow disease in bovines. These neurological conditions are caused by prions, which are infectious misfolded proteins. Could the microorganisms found in the Martian cave behave the same way? Probably not exactly, since prions are non-living. But then again, the clinical presentation in Jango and Mike was not exactly the same as in the prion diseases, so perhaps a Martian infection would have the unique characteristics described in the story. By the way, the BSL-4 containment facilities referred to by the crewmembers actually exist, and the National Research Council has recommended that a Mars Quarantine Facility should have BSL-4 level capability [58].

The notion of panspermia is an interesting one. It proposes that life is present throughout the universe and can be distributed from place to place by meteoroids, asteroids, and other moving heavenly bodies, including debris ejected into space by the collision of two bodies. It assumes that life forms can be trapped in a dormant state that can survive the effects of space travel for long periods of time until chancing upon the surface of another body with the right conditions to activate them and allow them to live and evolve in their new home.

Is there any evidence for such a notion? We know that some traveling bodies, such as comets, contain water, and it is thought that cometary collisions with planetary bodies may be one way that water accumulates on their surfaces. We also know that certain extremophile bacteria can exist on Earth in very inhospitable conditions involving high or low temperatures, highly acidic or alkaline conditions, high or low pressures, and even high radiation. Extremophiles have been found under the Antarctic ice, around volcanic vents and hot springs, and attached to rocks. Some can produce methane. Also, a study of the magnetic properties of a meteorite found on Earth (named ALH84001) that is thought to originate on Mars suggests that the interior was not heated to more than 40°C in its transit to Earth, which is well within the temperature range of some microorganisms [59]. This rock also contains carbonate globules rimmed with iron-oxide particles that some have argued are fossils of ancient Martian bacteria. Finally, living bacterial spores have been retrieved from orbiting satellites, where they spent more than five years basking in ultraviolet light and a deep vacuum. Bacteria are very hardy microorganisms indeed! So who knows, maybe our ancestors originated on Mars after all.

8 References

1. Kanas, N., Manzey, D.: Space Psychology and Psychiatry, 2nd edn. Microcosm Press, El Segundo and Springer, Dordrecht (2008)
2. Kanas, N., Feddersen, W.: Behavioral, Psychiatric and Sociological Problems of Long-Duration Space Missions. NASA TM X-58067. National Aeronautics and Space Administration Manned Spacecraft Center, Texas (1971)
3. Bechtel, R.B., Berning, A.: The third-quarter phenomenon: Do people experience discomfort after stress has passed? In: Harrison, A.A., Clearwater, Y.A., McKay, C.P. (eds.) From Antarctica to Outer Space. Springer, New York (1991)
4. Kanas, N., Salnitskiy, V., Weiss, D.S., Grund, E.M., Gushin, V., Kozerenko, O., Sled, A., Bostrom, A., Marmar, C.R.: Crewmember and ground personnel interactions over time during shuttle/mir space missions. Aviat. Space Environ. Med. **72**, 453–461 (2001)
5. Kanas, N.A., Salnitskiy, V.P., Boyd, J.E., Gushin, V.I., Weiss, D.S., Saylor, S.A., Kozerenko, O.P., Marmar, C.R.: Crewmember and mission control personnel interactions during international station missions. Aviat. Space Environ. Med. **78**, 601–607 (2007)
6. Kelly, A.D., Kanas, N.: Communication between space crews and ground personnel: a survey of astronauts and cosmonauts. Aviat. Space Environ. Med. **64**, 795–800 (1993)
7. Kelly, A.D., Kanas, N.: Crewmember communication in space: a survey of astronauts and cosmonauts. Aviat. Space Environ. Med. **63**, 721–726 (1992)
8. Stuster, J.: Behavioral issues associated with long-duration space expeditions: review and analysis of astronaut journals. Experiment 01-E104 (Journals): Final Report. NASA/TM-2010-216130. Houston, Texas: NASA/Johnson Space Center (2010)
9. Kanas, N., Salnitskiy, V., Grund, E.M., Gushin, V., Weiss, D.S., Kozerenko, O., Sled, A., Marmar, C.R.: Interpersonal and cultural issues involving crews and ground personnel during shuttle/mir space missions. Aviat. Space Environ. Med. **71** (9), A11–16 (2000)

10. Boyd, J.E., Kanas, N.A., Salnitskiy, V.P., Gushin, V.I., Saylor, S.A., Weiss, D.S., Marmar, C.R.: Cultural differences in crewmembers and Mission Control personnel during two space station programs. Aviat. Space Environ. Med. **80**, 1–9 (2009)

11. Tomi, L., Kealey, D., Lange, M., Stefanowska, P., Doyle, V.: Cross-cultural training requirements for long-duration space missions: results of a survey of international space station astronauts and ground support personnel. Paper delivered at the Human Interactions in Space Symposium, May 21, 2007, Beijing, China (2007)

12. Sandal, G.M., Manzey, D.: Cross-cultural issues in space operations: a survey study among ground personnel of the European space agency. Acta Astronaut. **65**, 1520–1529 (2009)

13. Nechaev, A.P., Polyakov, V.V., Morukov, B.V.: Martian manned mission: what cosmonauts think about this. Acta Astronaut. **60**, 351–353 (2007)

14. Kanas, N., Saylor, S., Harris, M., Neylan, T., Boyd, J., Weiss, D.S., Baskin, P., Cook, C., Marmar, C.: High vs. low crewmember autonomy in space simulation environments. Acta Astronaut. **67**, 731–738 (2010)

15. Kanas, N., Harris, M., Neylan, T., Boyd, J., Weiss, D.S., Cook, C., Saylor, S.: High vs. low crewmember autonomy during a 105-day Mars simulation mission. Acta Astronaut. **69**, 240–244 (2011)

16. Roma, P.C., Hursh, S.R., Hienz, R.D., Brinson, Z.S., Gasior, E.D., Brady, J.V.: Interactive effects of autonomous operations and circadian factors on crew performance, behavior, and stress physiology. Paper # IAC-12-A1.1.11. International Astronautical Federation. Proceedings, 63th International Astronautical Congress, Naples, Italy, 1–5 October 2012 (2012)

17. Astronaut tells of down side to space life, The New York Times. January 22, 1997. http://www.nytimes.com/1997/01/22/us/astronaut-tells-of-down-side-to-space-life.html?ref=johneblaha (1997)

18. Kanas, N., Salnitskiy, V., Gushin, V., Weiss, D.S., Grund, E.M., Flynn, C., Kozerenko, O., Sled, A., Marmar, C.R.: Asthenia: does it exist in space? Psychosom Med. **63**, 874–880 (2001)

19. Boyd, J.E., Kanas, N., Gushin, V.I., Saylor, S.: Cultural differences in patterns of mood states on board the international space station. Acta Astronaut. **61**, 668–671 (2007)

20. Aldrin, E.E.: Return to Earth. Random House, New York (1973)

21. Isay, R.A.: The submariners' wives syndrome. Psychiatric Quart. **42**, 647–652 (1968)

22. Pearlman Jr., C.A., Separation reactions of married women. Am. J. Psychiatry. **126**, 946–950 (1970)

23. Lebedev, V.: Diary of a Cosmonaut: 211 Days in Space. College Station. Phytoresource Research Information Service, Texas (1988)

24. Linenger, J. M.: Off the Planet: Surviving Five Perilous Months aboard the Space Station Mir. McGraw-Hill, New York (2000)

25. Belew, L.F.: Skylab: Our First Space Station. NASA SP-400. National Aeronautics and Space Administration, Washington D.C. (1977)

26. Cooper Jr., H.S.F.: A House in Space. Holt, Rhinehart & Winston, New York (1976)
27. Gushin, V.I., Zaprisa, N.S., Kolinitchenko, T.B., Efimov, V.A., Smirnova, T.M., Vinokhodova, A.G., Kanas, N.: Content analysis of the crew communication with external communicants under prolonged isolation. Aviat. Space Environ. Med. **68**, 1093–1098 (1997)
28. Kanas, N., Sandal, G., Boyd, J.E., Gushin, V.I., Manzey, D., et al.: Psychology and culture during long-duration space missions. Acta Astronaut. **64**, 659–677 (2009)
29. Gushin. V.I., Pustynnikova, J.M., Smirnova, T.M.: Interrelations between the small isolated groups with homogeneous and heterogeneous composition. Human Perf. in Extreme Environ. **6**, 26–33 (2001)
30. Gushin, V.I., Efimov, V.A., Smirnova, T.M., Vinokhodova, A.G, Kanas, N.: Subject's perception of the crew interaction dynamics under prolonged isolation. Aviat. Space Environ. Med. **69**, 556–561(1998)
31. MacCallum, T., Poynter, J., Bearden, D.: Lessons learned from Biosphere 2: When viewed as a ground simulation/analog for long duration human space exploration and settlement. Paper # 2004-01-2473. SAE International. http://www.jane-pounter.com/ documents/LessonsfromBio2 (2004)
32. Urbina, D., Charles, R.: Enduring the isolation of interplanetary travel: A personal account of the Mars 500 mission. Paper # IAC-12-A1.1.1. International Astronautical Federation. Proceedings, 63th International Astronautical Congress, Naples, Italy, 1–5 Oct 2012 (2012)
33. Gushin, V., Shved, D., Ehmann, B., Balazss, L., Komarevtsev, S.: Crew-MC interactions during communication delay in Mars-500. Paper # IAC-12-A1.1.2. International Astronautical Federation. Proceedings, 63th International Astronautical Congress, Naples, Italy, 1–5 Oct 2012 (2012)
34. Sandal, G.M.: "Groupthink" on a mission to Mars: Results from a 520 days space simulation study. Paper # IAC-12-A1.1.3. International Astronautical Federation. Proceedings, 63th International Astronautical Congress, Naples, Italy, 1–5 Oct 2012 (2012)
35. Van Baarsen, B., Ferlazzo, F., Ferravante, D., Smit, J., van der Pligt, J., van Duijn, M.: The effects of extreme isolation on loneliness and cognitive control processes: Analyses of the Lodgead data obtained during the Mars105 and the Mars520 studies. Paper # IAC-12-A1.1.4. International Astronautical Federation. Proceedings, 63th International Astronautical Congress, Naples, Italy, 1–5 Oct 2012 (2012)
36. Van Baarsen, B.: Person autonomy and voluntariness as important factors in motivation, decision making, and astronaut safety: first results from the Mars500 LODGEAD study. Acta Astronaut. **87**, 139–146 (2013)
37. Basner, M., Dinges, D., Mollicone, D., Savelev, I., Ecker, A., Di Antonio, A., Jones, C., Hyder, E., Kan, K., Morukov, B., Sutton, J.: Behaviour, performance and psychosocial issues in space. Paper # IAC-12-A1.1.5. International Astronautical Federation. Proceedings, 63th International Astronautical Congress, Naples, Italy, 1–5 Oct 2012 (2012)

38. Basner, M., Dinges, D.F., Mollicone, D., Ecker, A., Jones, C.W., Hyder, E.C., Antonio, A.D., Savelev, I., Kan, K., Goel, N., Morukov, B.V., Sutton, J.P.: Mars 520-d mission simulation reveals protracted crew hypokinesis and alterations of sleep duration and timing. Proc. Nat. Acad. Sci. **110**(7), 2635–2640 (2013)

39. Tafforin, C.: The Mars-500 crew in daily life activities: Ethological study. Paper # IAC-12-A1.1.6. International Astronautical Federation. Proceedings, 63th International Astronautical Congress, Naples, Italy, 1–5 Oct 2012 (2012)

40. Woodmansee, L.S.: Sex in Space. CG Publishing, Burlington (2006)

41. Committee for the Decadal Survey on Biological and Physical Sciences in Space: Recapturing a Future for Space Exploration. Space Studies Board, National Research Council. The National Academies Press, Washington D.C. (2011)

42. Clement, G.: Fundamentals of Space Medicine. Microcosm Press, El Segundo Springer, Dordrecht (2005)

43. Buckey Jr., J.C.: Space Physiology. Oxford University Press, New York (2006)

44. Karash, Y.: Sex in space: From Russia…with love, Space.com. http://web.archive. org/web/20080710053017/http://www.space.com/scienceastronomy/generalscience/russian_sex_studies_000316.html Accessed 16 Mar 2000 (2000)

45. Wall, M.: Russian official says there's been no sex in space…yet, NBCNews. com. http://www.nbcnews.com/id/42731409/ns/technology_and_science-space/ Accessed 23 April 2011 (2011)

46. Ball, J.R., Evans Jr, C.H. (eds.): Safe Passage: Astronaut Care for Exploration Missions. Committee on Creating a Vision for Space Medicine during Travel Beyond Earth Orbit. Institute of Medicine. National Academy Press, Washington D.C. (2001)

47. Mullane, M.: Riding Rockets: The Outrageous Tales of a Space Shuttle Astronaut. Scribner, New York (2006)

48. Boyle, A. Outer-space sex carries complications, MSNBC.com. http://www. nbcnews.com/id/14002908/ns/technology_and_science-space/t/outer-space-sex-carries-complications/ Accessed 24 July 2006 (2006)

49. Palinkas, L.A.: Group adaptation and individual adjustment in antarctica: a summary of recent research. In: Harrison A.A., Clearwater Y.A., McKay C.P. (eds.) From Antarctica to Outer Space. Springer, New York (1991)

50. Suedfeld, P.: Homo invictus: The indomitable species. Canadian Psychol. **38**, 164–173 (1998)

51. Suedfeld, P.: Applying positive psychology in the study of extreme environments. Human Perf. in Extreme Environ. **6**, 21–25 (2001)

52. Ihle, E.C., Ritsher, J.B., Kanas, N.: Positive psychological outcomes of space flight: an empirical study. Aviat. Space Environ. Med. **77**, 93–101 (2006)

53. Suedfeld, P., Legkaia, K., Brcic, J.: Changes in the hierarchy of value references associated with flying in space. J. Personality **78**(5), 1–25 (2010)

54. Methane found in desert soils bolsters theories that life could exist on Mars. Science-Daily. 8 Nov 2005. http://www.sciencedaily.com/releases/2005/11/051107083842. htm (2005)

55. Diaz, B., Schulze-Makuch, D.: Microbial survival rates of *Escherichia coli* and *Deinococcus radiodurans* under low temperature, low pressure, and UV-irradiation conditions, and their relevance to possible Martian life. Astrobiology **6**(2), 332–347 (2006)
56. Model methanogens provide clues to possible Mars life. ScienceDaily. 28 May 2007. http://www.sciencedaily.com/releases/2007/05/070525204839.htm (2007)
57. De Vera, J-P. P., Schulze-Makuch, D., Khan, A., Lorek, A., Koncz, A., Mohlmann, D., Spohn, T.: The adaptation potential of extremophiles to Martian surface conditions and its implication for the habitability of Mars. Geophys. Res. Abstr. **14**:EGU2012–2113 (2012)
58. Committee on Planetary and Lunar Exploration, Space Studies Board, Division on Engineering and Physical Sciences, National Research Council: The Quarantine and Certification of Martian Samples. National Academy Press, Washington D.C. (2002)
59. Weiss, B.P., Kirschvink, J.L.: Life from space? Testing panspermia with Martian meteorite ALH84001. The Planetary Report, November/December, 8–11 (2000)

Printed by Printforce, the Netherlands